Confidence Intervals for Discrete Data in Clinical Research

Chapman & Hall/CRC Biostatistics Series

Recently Published Titles

Mathematical and Statistical Skills in the Biopharmaceutical Industry: A Pragmatic Approach
Arkadiy Pitman, Oleksandr Sverdlov, L. Bruce Pearce

Real-World Evidence in Drug Development and Evaluation
Harry Yang, Binbing Yu

Cure Models: Methods, Applications, and Implementation
Yingwei Peng, Binbing Yu

Bayesian Analysis of Infectious Diseases
COVID-19 and Beyond
Lyle D. Broemeling

Statistical Meta-Analysis using R and Stata, Second Edition
Ding-Geng (Din) Chen and Karl E. Peace

Advanced Survival Models
Catherine Legrand

Structural Equation Modeling for Health and Medicine
Douglas Gunzler, Adam Perzynski and Adam C. Carle

Signal Detection for Medical Scientists
Likelihood Ratio Test-based Methodology
Ram Tiwari, Jyoti Zalkikar, and Lan Huang

Single-Arm Phase II Survival Trial Design
Jianrong Wu

Methodologies in Biosimilar Product Development
Sang Joon Lee, Shein-Chung Chow (eds.)

Statistical Design and Analysis of Clinical Trials
Principles and Methods, Second Edition
Weichung Joe Shih, Joseph Aisner

Confidence Intervals for Discrete Data in Clinical Research
Vivek Pradhan, Ashis K. Gangopadhyay, Sandeep M. Menon, Cynthia Basu, and Tathagata Banerjee

For more information about this series, please visit: https://www.routledge.com/Chapman--Hall-CRC-Biostatistics-Series/book-series/CHBIOSTATIS

Confidence Intervals for Discrete Data in Clinical Research

Vivek Pradhan
Ashis K. Gangopadhyay
Sandeep M. Menon
Cynthia Basu
Tathagata Banerjee

CRC Press
Taylor & Francis Group
Boca Raton London New York

CRC Press is an imprint of the
Taylor & Francis Group, an **informa** business
A CHAPMAN & HALL BOOK

First edition published 2022
by CRC Press
6000 Broken Sound Parkway NW, Suite 300, Boca Raton, FL 33487-2742

and by CRC Press
2 Park Square, Milton Park, Abingdon, Oxon, OX14 4RN

CRC Press is an imprint of Taylor & Francis Group, LLC

Library of Congress Cataloging-in-Publication Data

ISBN: 9781138048980 (hbk)
ISBN: 9781032128634 (pbk)
ISBN: 9781315169859 (ebk)

DOI: 10.1201/9781315169859

Typeset in CMR10 font
by KnowledgeWorks Global Ltd.

Contents

Preface

This book is intended for students and professionals involved in biomedical research. The book arose from the experiences of the authors, who have been involved with biomedical research and worked with biomedical data for many years. The book is designed to address some fundamental questions in statistical inference, such as the best practices in analyzing discrete data and how to implement the analyses using statistical software. Of course, the answers to these questions are available in literature and software manuals, and can be found as long as the user is willing to undertake a long and arduous journey through the forest of scattered and often contradictory information. The authors, over many years, have undergone the process of identifying, evaluating, implementing, and teaching wide-ranging methodologies related to inferential questions that commonly occur in clinical research. This book is the culmination of their collective experiences in the field and is written in a cookbook style to alleviate the frustrations that many practitioners feel in choosing and implanting optimal data analysis methods to answer inferential questions quickly and efficiently.

Therefore, the target audience of the book are:

- The practitioners who wish to learn about traditional as well as cutting-edge data analysis tools and their implementations for discrete data without unduly worrying about the theoretical foundations.
- The students in biomedical fields who want to familiarize themselves with data analysis methods without being distracted by the theoretical underpinnings of these methods.
- Instructors of intermediate-level data analysis courses who wish to teach reliable and robust approaches to statistical analysis of discrete data.

The pedagogical approach of the book can be summarized as follows. Each chapter of the book:

- Starts with a brief introduction to the inferential question being addressed.
- Reviews the common approaches to the inferential questions and discusses the limitations of the methods.

- Introduces alternative methods from the literature that can be used to address such limitations.
- Provides comparisons of the methods.
- Concludes with a discussion and recommendation of inferential methods under various scenarios.

The book has been written with four overriding principles in mind. The first is based on the authors' belief that in clinical research, there is an over-reliance on hypothesis testing and p-value. The significance of a hypothesis testing procedure based on an artificial cut-off of a p-value (usually 0.05) often highlighted on statistical software with symbols ***,**, etc., creates an illusion of simplicity and robs the researcher of a nuanced understanding of the evidence embedded in the data. On the other hand, a confidence interval allows the user to recognize the uncertainties of the inference, but at the same time, can be used to evaluate a hypothesis by leveraging the duality between the two inferential tools. Therefore, the book focuses on the exposition of confidence interval procedures as a data analytic tool but at the same time charts out a clear pathway for the users who need the "clarity" of the p-value.

The second guiding principle of the book is to introduce the methodologies with minimal theory but at the same time providing some perspective so that it is clear to a reader the relevance of the methods in the broader context. In this day and age, it is easy for a user to analyze data using software, but the hard part to recognize the most appropriate tool in the buffet of methods available in the software. Therefore, each chapter in the book lays out the methodological options available to the user. Without going into analytical details, it offers insights into the advantages and disadvantages of the methods. Although in this book, many technical details are intentionally omitted for the sake of simplicity, the users interested in learning about the theoretical foundations of the methods are referred to resources where such information is available.

The third guiding principle of the book is to integrate the computational tools seamlessly with all the methodologies. The statistical methods introduced in the book have been illustrated primarily with software packages SAS and StatXact; to a lesser extent, R. SAS and StatXact are the two widely used software platforms in clinical research. The focus on these computational tools should be immensely beneficial to the target audience of this book. However, some of the procedures discussed in the book do not always have ready-made codes available in these software packages. On those occasions, detailed documented codes have been developed for this book and will be made available to the users via the book's resource website.

The fourth guiding principle is to ensure that the book's focus remains on applications in clinical research and provides direct benefit to the users involved in the field. The methods discussed in the book are illustrated with examples that reflect the long experience of the authors in the area of clinical research and drug regulatory regime. The applications are designed to provide concrete illustrations of the analytic approaches consistent with regulatory guidelines of clinical research. Therefore, the book covers classical frequentist methodologies that, over many years, have been widely used in clinical research, but at the same time explores Bayesian methods that have in recent years received increased acceptance in the FDA regulatory process.

The book begins with a brief review of fundamental ideas and notational conventions of inferential statistics, presented in an intuitive and non-technical manner. The second chapter of the book offers a critic of the overreliance on the p-value in traditional inferential procedures. Chapters 3–6 introduce methodologies for binary outcomes covering inferences of proportions and a range of functionals of proportions useful in analyzing clinical data. The book concludes with Chapter 7, which presents methods for the analysis of count data.

The material presented in the book is intended for a broad range of audiences in mind. The methodologies discussed can be reviewed in an à la carte approach so that the book can be used as a reference. A researcher can quickly jump into a specific section of the book to find the relevant information he or she is seeking without reading the whole book. However, one can use the book for a comprehensive understanding of the methodologies by going over the chapters in the order they are presented. For example, the book can be an ideal supplement for a course on statistics for clinical research. Ultimately, the authors hope that the book will be helpful to clinical researchers in making informed choices in commonly used data analytic tools.

Authors

Vivek Pradhan has been working in the industry for more than 20 years. Currently, he is a Senior Director of Statistics in Early Clinical Development at Pfizer Inc. where he is responsible for managing all the statistical aspects of drug development from pre-clinical to Phase IIB trials. He has been publishing methodological papers on discrete data, and is a regularly invited speaker in several industry conferences and forums.

Ashis K. Gangopadhyay is an associate professor of statistics in the Department of Mathematics and Statistics at Boston University. His research areas include predictive modeling in clinical research, nonparametric and semiparametric methods, and analysis of financial data. He has authored numerous extensively cited research papers and mentored many PhD students.

Sandeep M. Menon is Senior Vice President and Head of Early Clinical Development at Pfizer Inc. and holds adjunct faculty positions at Boston University School of Public Health, Tufts University School of Medicine, and the Indian Institute of Management. At Pfizer, he is in the Worldwide Research, Development and Medical Leadership Team, and leads a multifunctional global team. Before joining the industry, he practiced medicine in Mumbai and was Resident Medical Officer. Dr. Menon is an elected fellow of the American Statistical Association (ASA), was awarded the Young Scientist Award by the International Indian Statistical Association, the Statistical Excellence Award in Pharmaceutical Industry by Royal Statistical Society, UK, and, recently, the Distinguished Alumni Award by the Department of Biostatistics at Boston University School of Public Health. He earned his MD from Karnataka University, India and later completed his master's in epidemiology and biostatistics, PhD in biostatistics at Boston University, and research assistantship at Harvard Clinical Research Institute. He has published more than 50 scientific original publications and book chapters and co-authored/co-edited 6 books.

Cynthia Basu has been involved in research in clinical trials and Bayesian methods. She is currently working as an Associate Director in Statistics in Early Clinical Development at Pfizer Inc. where she works on early phase trials in oncology. Her research interests include topics in clinical trial designs, Bayesian methods, adaptive trials, and historical borrowing.

Tathagata Banerjee has been engaged in teaching and research in statistics for more than three decades. Currently, he is a professor at the Indian Institute of Management Ahmedabad, India. His research interest is primarily focussed on developing statistical methodologies for drawing inference from different kinds of data. His research is published regularly in peer-reviewed journals, and he has given lectures and taught in various universities across the world.

1

A Brief Review of Statistical Inference

1.1 Introduction

"I believe sanity and realism can be restored to the teaching of Mathematical Statistics most easily and directly by entrusting such teaching largely to men and women who have had personal experience of research in the Natural Sciences." – Sir R. A. Fisher

Statistics is the science of understanding uncertainty and variability of data. This is a challenging undertaking, and over many decades, several schools of thought have emerged to develop methodologies to achieve this goal. On the one hand, the diversity of these approaches has enriched the toolbox of statisticians that they can rely on in developing strategies to understand the nature of uncertainty. Still, on the other, it has added complexities to various modeling and inferential procedures. This necessitates critical evaluations of multiple approaches to solve a common set of problems and develop simplified recommendations that a practicing statistician can rely on. This book's primary aim is to explore various statistical solutions to the questions that arise in the context of the analysis of discrete data and provide guidance in their usage in practical applications in clinical research.

Many scientific questions start with a hypothesis, and scientific progress relies on evaluating those hypotheses in light of evidence embedded in data. A hypothesis is formulated as a consequence of specific facts that have been observed and is designed to make broad generalizations of the facts. In that sense, the inferential procedures in statistics follow inductive logic. However, such inferential methods may be based on only the evidence in the data, or it may utilize additional subjective information available to the researcher. The difficulty with this approach is that the inductive reasoning process increases uncertainties, which lends itself to the subjective interpretation of the data.

In general, there are two main strands of inferential approaches, namely frequentist and Bayesian. The frequentist (or objective) approach characterizes uncertainty as a long-run frequency of outcomes observed in very large samples. Therefore, the uncertainty of inference is characterized by the idea of repeated drawing of hypothetical samples from a population, each generating a specific approximation of the parameter of interest. The distribution of such values, called the sampling distribution, provides a way to describe the

DOI: 10.1201/9781315169859-1

inherent uncertainty of an inference. On the other hand, the Bayesian framework allows certain subjective information to be incorporated into the inference machinery. This is achieved by assigning a probability distribution, known as a prior distribution, on the parameter space. Therefore, a prior distribution describes the plausible values of a parameter without accounting for the data. The probability mechanism, specifically the Bayes rule, is utilized to update the parameter's likely values by accounting for the observed data. The prior information can result from specific knowledge of the parameter, or it can be chosen, as is the case often in practice, with computational convenience in mind.

The debate on the superiority between frequentist and Bayesian approaches is old and stale. The two inferential approaches provide a different worldview when it comes to statistics applications, and it shows that there are foundational questions in statistics that remain unsettled. However, in practical terms, both approaches allow different ways of looking at information, and any practicing statistician would benefit from the differing interpretation of data afforded by these methods. In this book, we have discussed both frequentist and Bayesian methods, and where applicable, given our thoughts on the suitability of these approaches within the context of the problem.

1.2 The frequentist approach

Consider a random sample $X_1, X_2, \ldots, X_n$ from statistical model given by the probability density function $f(x;\theta)$, where θ is the parameter of interest. One key inferential question is how to estimate the parameter θ based on the sample data. There are diverse frequentist approaches to estimate the parameter θ. The likelihood-based approach advocated by R. A. Fisher allows an elegant way to resolve inferential problems about a parameter by answering the following question: What value(s) of the parameter θ is most likely to have generated the data that we have observed? To understand this a bit better, consider a sequence of $n = 10$ Bernoulli trials that resulted in $x = 2$ successes. If θ is the true probability of success, then the likelihood of θ is given by

$$L(\theta) = P_\theta(X = 2) \propto \theta^2(1-\theta)^8 \tag{1.1}$$

The likelihood function given in Figure(1.1) shows that the likelihood is maximized at $\hat{\theta} = \frac{2}{10} = 0.2$, which is the sample proportion and the maximum likelihood estimate of the parameter θ.

In general, for independent samples, the maximum likelihood estimate (MLE) is given by the value of the parameter $\hat{\theta}$ that maximizes the likelihood function

$$L(\theta) = \prod_{i=1}^{n} f(x_i;\theta) \tag{1.2}$$

FIGURE 1.1
The likelihood function of the true probability of success θ in Bernoulli trails for $n = 10$ and $x = 2$.

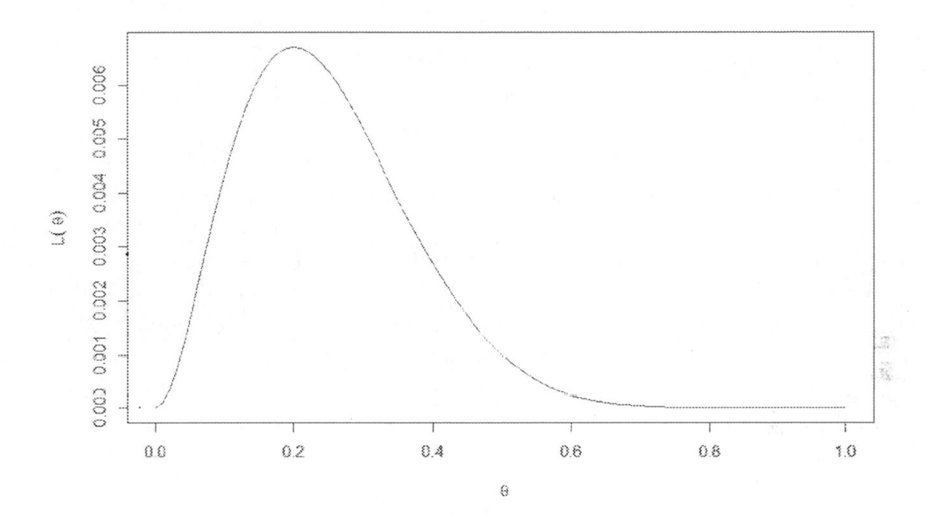

Equivalently, $\hat{\theta}$ maximizes the log of the likelihood function

$$logL(\theta) = \sum_{i=1}^{n} logf(X_i; \theta) \tag{1.3}$$

For Binomial experiment, with n independent trails and x successes, the log-likelihood function is given by

$$logL(\theta) = xlog\theta + (n - x)log(1 - \theta) \tag{1.4}$$

which is maximized at the sample proportion $\hat{\theta} = \frac{x}{n}$, the MLE of θ.

Regardless of how the parameter θ is estimated, whether it is by maximum likelihood or some other method, the frequentist approach to the statistical inference of parameter θ is derived from the probability distribution of the estimate $\hat{\theta}$, which is the sampling distribution. The distribution of $\hat{\theta}$ may be known for any sample size, called the exact sampling distribution. For example, if we have independent samples $X_1, X_2, \ldots, X_n$ from a normal population with mean μ and variance σ^2, the MLE of μ is the sample mean $\bar{X}$, and the exact sampling distribution of $\bar{X}$ is given by

$$\bar{X} \sim N\left(\mu, \frac{\sigma^2}{n}\right) \tag{1.5}$$

Similarly, the MLE of σ^2 is $\hat{\sigma}^2 = \frac{1}{n}\sum_{i=1}^{n}(X_i - \bar{X})^2$, but the commonly used unbiased estimator of σ^2 is $s^2 = \frac{n}{n-1}\hat{\sigma}^2 = \frac{1}{n-1}\sum_{i=1}^{n}(X_i - \bar{X})^2$. Under the assumption that the population is normally distributed, the exact sampling distribution of s^2 is given by

$$\frac{(n-1)s^2}{\sigma^2} \sim \chi^2_{n-1} \tag{1.6}$$

where χ^2_{n-1} is the chi-square distribution with $(n-1)$ degrees of freedom.

However, more often than not, the exact sampling distribution of an estimator $\hat{\theta}$ is not known. In that case, we rely on the asymptotic distribution of $\hat{\theta}$, i.e., the approximate probability distribution of $\hat{\theta}$ when the sample size is sufficiently large. A well-known example of asymptotic distribution is the Central Limit Theorem (CLT). Suppose $X_1, X_2, \ldots, X_n$ is a random sample from a population, not necessarily normal, with mean μ and variance σ^2, then for a sufficiently large sample size n, the sampling distribution of $\bar{X} \sim N(\mu, \frac{\sigma^2}{n})$, the same as the exact distribution. In general, $n \geq 30$ can be considered a sufficiently large sample. In this regard, the MLE has a significant advantage. It can be shown that under certain conditions on the true population distribution $f(x;\theta)$ characterized by a real parameter θ, the sampling distribution of the MLE $\hat{\theta}$ is given by

$$\hat{\theta} \sim N\left(\theta, \frac{1}{nI(\theta)}\right) \tag{1.7}$$

$I(\theta)$ is called the Fisher Information, and is given by

$$I(\theta) = Var_{\theta}\left(\frac{\partial log f(x;\theta)}{\partial\theta}\right) \tag{1.8}$$

For Bernoulli experiment with $f(x;\theta) = \theta^x(1-\theta)^{1-x}$, $for\, x = 0, 1$, it is easy to see that $I(\theta) = \frac{1}{\theta(1-\theta)}$. Therefore, the corresponding asymptotic distribution of the MLE of θ is given by

$$\hat{\theta} \sim N\left(\theta, \frac{\theta(1-\theta)}{n}\right) \tag{1.9}$$

How good is the normal approximation? The answer depends on the sample size and the true value of the population proportion θ in the interval $(0, 1)$. We will discuss this issue in more detail in subsequent chapters.

1.2.1 Confidence interval methods

The two primary inferential procedures for a population parameter are estimation and hypothesis testing. In the estimation procedure, the objective is to utilize the sample data to develop the best possible approximation of the population parameter. A point estimator is a single value (or a point on the real line) that provides the best approximation of the parameter in some sense. For example, the sample proportion $\hat{\theta}$ is a point estimate of the population proportion, and it is "best" in the sense that it is the MLE, and also the UMVUE (Uniformly Minimum Variance Unbiased Estimator).

However, a more effective approach to estimating a population parameter is to use a confidence interval. As the name suggests, the idea is to come up with an interval, let's say $C(x) = [L(x), U(x)]$, with boundary points $L(x)$ and $U(x)$ determined by the sample data so that the true population parameter θ has a very high probability of being included in the interval. Therefore, the interval $C(x)$ contains all plausible values of the population parameter. The probability associated with the interval is the confidence coefficient, or when expressed as a percentage, is called the confidence level of the interval. The derivation of the confidence interval relies on the sampling distribution of the point estimate. For example, it follows from (1.9) that

$$P\left(-z_{\alpha/2} \leq \frac{\hat{\theta} - \theta}{\sqrt{\frac{\theta(1-\theta)}{n}}} \leq z_{\alpha/2}\right) = 1 - \alpha \tag{1.10}$$

where $z_{\alpha/2}$ is the $(1 - \alpha/2)$ percentile of the standard normal distribution, i.e., $P(Z < z_{\alpha/2}) = 1 - \alpha/2$. Replacing θ in the denominator by its estimate $\hat{\theta}$ and solving the inequality for θ, we get

$$P\left(\hat{\theta} - z_{\alpha/2}\sqrt{\frac{\hat{\theta}(1-\hat{\theta})}{n}} \leq \theta \leq \hat{\theta} + z_{\alpha/2}\sqrt{\frac{\hat{\theta}(1-\hat{\theta})}{n}}\right) = 1 - \alpha \tag{1.11}$$

Therefore, a $(1 - \alpha)$ confidence interval for θ is given by

$$\left[\hat{\theta} - z_{\alpha/2}\sqrt{\frac{\hat{\theta}(1-\hat{\theta})}{n}}, \hat{\theta} + z_{\alpha/2}\sqrt{\frac{\hat{\theta}(1-\hat{\theta})}{n}}\right] \tag{1.12}$$

which is known as the Wald's confidence interval.

The coverage probability of a confidence interval $C(x)$ is defined as

$$\text{Coverage Probability} = P_\theta(\theta \in C(x)) \tag{1.13}$$

i.e., the probability that the true value of the population parameter is "covered" by the interval. Another useful way to think about the coverage probability is in terms of the expected value of the indicator function that the interval contains the true parameter θ. In particular, in the context of estimating a binomial proportion θ, we can express the coverage probability as

$$\text{Coverage Probability} = \sum_{x=0}^{n} I(\theta, x)\binom{n}{x}\theta^x(1-\theta)^{n-x} \tag{1.14}$$

where $I(\theta, x)$ is the indicator function that takes the value 1 if $\theta \in C(x)$ and takes the value 0 if the interval $C(x)$ does not contain θ.

How is the coverage probability related to the confidence coefficient of the interval? The confidence coefficient is the minimum coverage probability for all possible θ in the parameter space. More specifically,

$$\text{Confidence Coefficient} = \inf_{\theta} P_{\theta}(\theta \in C(x)) \tag{1.15}$$

where inf is the *infimum*, i.e., roughly speaking, is the minimum of the coverage probabilities. In general, the coverage probability is the same as the confidence coefficient, but it is not necessarily the case. In particular, when a confidence interval is derived from an approximate sampling distribution such as (1.9), the coverage probability may depend on the true value of the parameter θ, especially for small to moderate sample sizes. In that case, the confidence coefficient can be significantly lower than the theoretical value of $(1-\alpha)$. This is a critical question in this context, and in this book, we address this issue in detail.

It is also worth mentioning that while coverage probability is the primary performance criterion of a confidence interval, there are secondary considerations that can play significant roles in identifying the "best" method. Of course, for a real-valued parameter, the narrowest interval is preferable among confidence intervals with comparable coverage probabilities. However, the length of a confidence interval is a random quantity. Therefore, it makes sense to consider the expected length instead, i.e., for a confidence interval $C(x)$ of parameter θ

$$\text{Expected Length} = E_{\theta}\,(length\,of\,C(X)) \tag{1.16}$$

where the expectation is with respect to the true probability model. If $C(x)$ is a confidence set in multiparameter problem, the expected length is replaced by the expected volume. Also, the definition of length depends on the parameter of interest. For example, if $C(x) = [L(x), U(x)]$ is the confidence interval for a population proportion θ, then clearly, $length\,of\,C(x) = U(x) - L(x)$. However, if the parameter of interest is *odds* $\eta = \frac{\theta}{1-\theta}$, then the notion of length of $C(x)$ can be modified to a more meaningful measure $(logU(x) - logL(x))$.

It should be intuitively clear that there is a connection between coverage probability and the expected length of a confidence interval. In particular, consider the false coverage probability of a confidence interval, which is the probability that the interval covers a *false* value θ^* when θ is the true value of the parameter. Specifically, for a two-sided confidence interval $C(x)$

$$\text{False Coverage Probability} = P_{\theta}\,(\theta^* \in C(x)) \tag{1.17}$$

It was shown by Ghosh [42] and Pratt [73] that the expected length of a confidence interval is the total (integrated) false coverage probability. Therefore, in essence, minimizing the false coverage probability achieves the same goal as minimizing the expected length of a confidence interval.

There is another consideration in evaluating the performance of a confidence interval worth mentioning, which is the location of the interval. Specifically, the mesial and distal directions of confidence intervals refer to the left and right directions of the interval relative to the true population parameter. Newcombe [64] introduced a measure of the location of a confidence interval as the ratio of the mesial non-coverage probability (MNCP) relative to the overall non-coverage probability NCP (1 − coverage probability) of the interval. A value in the range of 0.4 to 0.6 is considered satisfactory as it suggests a level of symmetry of the interval in the mesial and distal directions. Although this is an interesting and desirable property, it is a secondary consideration in judging the performance of a confidence interval.

It is possible that for a confidence interval $C(x) = [L(x), U(x)]$, either the lower limit $L(x)$ or the upper limit $U(x)$ to be ∞, and in that case, the confidence interval is one-sided. In particular, an interval of the form $(-\infty, U(x)]$ is an upper $(1-\alpha)$ confidence interval. Similarly, $[L(x), \infty)$ a lower $(1-\alpha)$ confidence interval. Note that $-\infty$ or ∞ are the placeholders for the smallest and largest parameter values in the parameter space. For example, in Bernoulli trails, from Equation 1.9, following the same approach as the two-sided confidence interval described above, a $(1-\alpha)$ upper confidence interval for the population proportion θ is given by $\left(0, \text{Min}\{\hat{\theta} + z_\alpha\sqrt{\frac{\hat{\theta}(1-\hat{\theta})}{n}}, 1\}\right]$, and a $(1-\alpha)$ lower confidence interval is given by $\left[\text{Max}\{\hat{\theta} - z_\alpha\sqrt{\frac{\hat{\theta}(1-\hat{\theta})}{n}}, 0\}, 1\right)$. There are many situations where a one-sided confidence interval is more appropriate than a two-sided one. For example, suppose a medical device manufacturer wants to estimate the rate of defective units produced in the manufacturing process. In that case, the manufacturer might be interested in an upper limit of the rate of the defective units, and hence an upper confidence interval would be the suitable approach.

1.2.2 Hypothesis testing methods

A confidence interval procedure attempts to answer the question: What values of the population parameter can plausibly generate the observed data? On the other hand, a hypothesis testing procedure asks: Is the sample data consistent with a hypothesized value of the population parameter? Therefore, a hypothesis testing question is a decision problem. The objective is to decide if a research hypothesis, called the alternative hypothesis and denoted by H_a, is true, or its complement, called the null hypothesis and denoted by H_0, is true. One can formulate the problem in the following manner: let θ be the parameter of interest, and it belongs to a parameter space Θ, and suppose the research hypothesis is that the true value of the parameter belongs to a subset of Θ, say Θ_1. Therefore, the null and the alternative hypothesis is given by

$$H_0 : \theta \in \Theta_0 \; vs \; H_a : \theta \in \Theta_1 \tag{1.18}$$

where $\Theta_0 = \Theta_1^c$, i.e., Θ_0 is the complement of Θ_1. Specifically, let θ be the success rate of a particular vaccine in producing antibodies at the desired level, and the research hypothesis is that the true success rate is greater than

95%. In that case, $\Theta = [0, 1]$, $\Theta_1 = (0.95, 1]$ and $\Theta_0 = [0, 0.95]$, and the null and the alternative hypotheses in (1.18) can be written as

$$H_0 : \theta \leq 0.95 \ vs \ H_a : \theta > 0.95 \tag{1.19}$$

The table below shows the two types of errors that can occur in a hypothesis testing problem.

	True State of Nature	
Decision	H_0 is true	H_a is true
Accept H_0	Correct decision	False negative (type II error)
Reject H_0	False positive (type I error)	Correct decision

Let α be the rate of false-positive (probability of type I error) and let β be the rate of false-negative (probability of type II error), Then $1 - \beta$, called the power of the test, is the probability of making a correct decision when the alternative hypothesis is true. In a decision problem, the objective is to develop a decision rule such that both α and β are small, i.e., a decision rule with a low false-positive rate with high power. The strategies to come up with such optimal decision rule is beyond the scope of this discussion. Still, interested readers can reference Casella and Berger (2002) for a relatively gentle introduction on this topic. However, referring to (1.19), it is clear that a large value of $\hat{\theta}$ would be more supportive of the alternative hypothesis, or equivalently, the decision rule should reject H_0 if

$$Z = \frac{\hat{\theta} - \theta_0}{\sqrt{\frac{\theta_0(1-\theta_0)}{n}}} > k \tag{1.20}$$

for some large constant k, called the critical value of the test, and Z is the test statistic. If the null hypothesis is true, then from (1.9), it follows that asymptotically $Z \sim N(0, 1)$. Therefore, $k = z_\alpha$ ensures that the probability of type I error, or the false positive rate, is α, which is called the level α test.

For example, in the vaccination trial discussed above, suppose the observed success rate is 96% in a random sample of $n = 2000$ vaccinated subjects, and we wish to test the hypothesis (1.19) at $\alpha = 0.01$ level. Therefore, the value of the test statistic $Z = (0.96 - 0.95)/\sqrt{\frac{0.95(1-0.95)}{2000}} \approx 2.05$. Since $z_{0.01} \approx 2.33$, there is insufficient evidence in the sample to reject the null hypothesis.

An equivalent approach to the decision rule of a hypothesis testing procedure is based on the p-value of a test. The p-value, or the observed significance level, is the probability of obtaining a value of the test statistic that is at least as favorable to the alternative hypothesis as the one observed from the sample data if the null hypothesis is true. Therefore, a small p-value suggests that if the null hypothesis is true, the observed value of the test static is unusual, and hence the alternative hypothesis should be preferred. A level α test rejects H_0 if the p-value $< \alpha$. In the vaccination example, the value of the test statistic is 2.05, and for the alternative hypothesis $H_a : \theta > 0.95$, a larger value of the test statistic would be favorable to the alternative hypothesis. Therefore, the p-value $= P(Z > 2.05) \approx 0.02$. Since the p-value $\not< 0.01$, we cannot reject H_0 for $\alpha = 0.01$, which is the same conclusion as above.

It is important to note that the confidence interval and hypothesis testing procedures are related, and there is a duality between the two approaches. We will not go into the details here but some critical facts are worth mentioning. The "duality" between confidence interval and hypothesis testing refers to the fact that the acceptance region of a $(1 - \alpha)$ level confidence interval can be inverted to obtain the corresponding α level test and vice-versa. Specifically, consider a two-sided test of proportion θ as $H_0 : \theta = \theta_0 \; vs \; H_a : \theta \neq \theta_0$, and the corresponding level-α rejection rule

$$Reject\, H_0\, if\, |Z| = \left|\frac{\hat{\theta} - \theta_0}{se(\hat{\theta})}\right| > z_{\alpha/2} \tag{1.21}$$

Therefore, we can define the acceptance region $A(\theta_0)$, i.e., the set of all estimates $\hat{\theta}$ for which the null hypothesis is accepted, as the complement of the rejection region. Hence, $A(\theta_0) = \{all\, \hat{\theta}\, for\, which\, |Z| \leq z_{\alpha/2}\}$, which implies

$$P(|Z| \leq z_{\alpha/2}) = 1 - \alpha \tag{1.22}$$

$$or, P\left(\left|\frac{\hat{\theta} - \theta_0}{se(\theta)}\right| \leq z_{\alpha/2}\right) = 1 - \alpha \tag{1.23}$$

$$or, P\left(\hat{\theta} - z_{\alpha/2} se(\hat{\theta}) \leq \theta_0 \leq \hat{\theta} + z_{\alpha/2} se(\hat{\theta})\right) = 1 - \alpha \tag{1.24}$$

Since θ_0 is an arbitrary value of θ, the last step gives a $(1 - \alpha)$ confidence interval of θ. Similarly, starting from a $(1 - \alpha)$ level confidence interval, one can generate a level α hypothesis testing procedure.

Figure (1.2) illustrates this duality between confidence interval and hypothesis testing. Suppose $\hat{\theta}*$ be the estimated proportion, then the interval $C(\hat{\theta}^*)$ is the confidence region. Similarly, for a fixed value of the parameter θ, say θ_0, $A(\theta_0)$ is the acceptance region of the null hypothesis $H_0 : \theta = \theta_0$. Finally, the duality holds for one-sided test and confidence interval as well. For example, consider the rejection rule for the one-sided alternative $H_a : \theta > \theta_0$ as

$$Reject\, H_0\, if\, Z = \frac{\hat{\theta} - \theta_0}{se(\hat{\theta})} > z_\alpha \tag{1.25}$$

Then, following the steps in (1.22), the $(1 - \alpha)$ lower confidence interval for θ is given by $[\hat{\theta} - z_\alpha se(\hat{\theta}), 1)$. Similarly, the $(1 - \alpha)$ upper confidence interval for θ given by $(0, \hat{\theta} + z_\alpha se(\hat{\theta})]$ can be obtained by inverting the lower-tail alternative hypothesis $H_a : \theta < \theta_0$.

FIGURE 1.2
Duality between confidence interval and hypothesis testing.

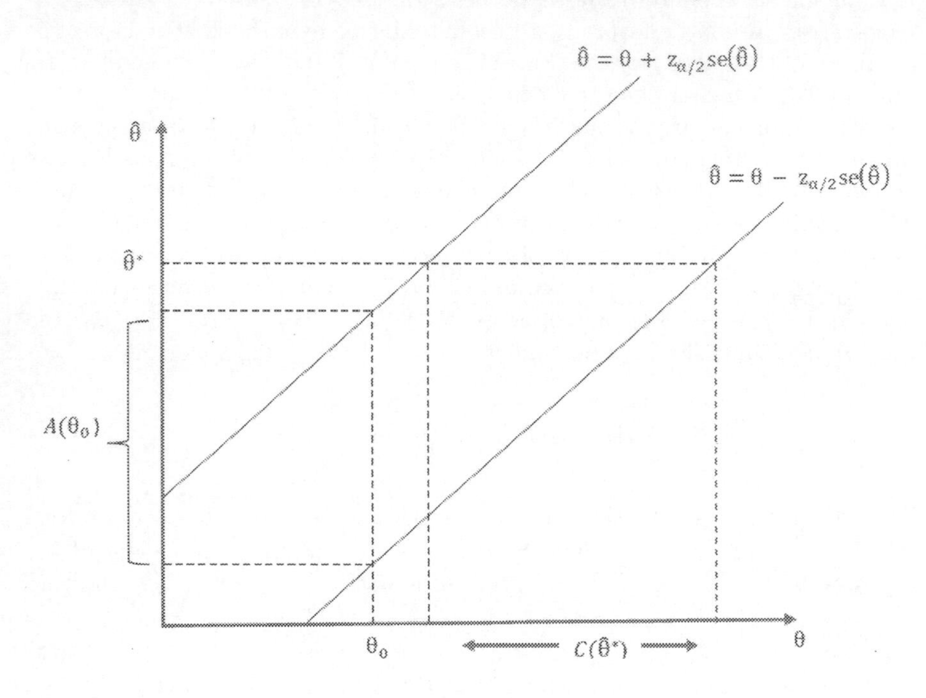

Therefore, confidence interval and hypothesis testing procedures are, in general, comparable, asking similar questions in two different ways. However, in many applications, a confidence interval is more informative and can provide a nuanced understanding of the parameter. In the next chapter, we will discuss the limitations of over-reliance on the p-value in a hypothesis testing procedure.

1.3 The Bayesian approach to inference

The Bayesian approach to inference relies on the Bayes rule in probability. It is a relatively straightforward concept than the frequentist approaches because inference about the population parameter is based on the probability model of the parameter of interest. However, the process of developing usable inference from sample data can be computationally challenging, even intractable. In recent years, Bayesian methods have become popular as new tools have been developed to address computational issues.

In the Bayesian framework, the uncertainty of a parameter θ is expressed as a probability model $\pi(\theta)$, called the prior distribution of θ, which reflects the prior belief of the experimenter regarding the parameter. The goal of the Bayesian approach is to use the Bayes rule in probability to update the prior distribution using the available data. The updated prior, called the posterior distribution, is then utilized to carry out the desired inferential procedure about the parameter θ. Here we will provide a brief review of Bayesian inference.

Suppose θ is the parameter of interest, and for simplicity, let's assume θ is a one-dimensional parameter. Let $\pi(\theta)$ be the prior distribution of θ, and the joint density of $\mathbf{x} = (x_1, x_2, \ldots, x_n)$ is given by $f(\mathbf{x}|\theta)$, then the posterior density, $g(\theta|\mathbf{x})$ is the conditional density of θ given $\mathbf{x}$, and can be calculated using the Bayes rule as follows:

$$g(\theta|\mathbf{x}) = \frac{\mathbf{f}(\mathbf{x}|\theta)\pi(\theta)\mathbf{d}\theta}{\int \mathbf{f}(\mathbf{x}|\theta)\pi(\theta)\mathbf{d}\theta} = \mathbf{cf}(\mathbf{x}|\theta)\pi(\theta), \tag{1.26}$$

where the denominator is the marginal density of $\mathbf{x}$ and is free of θ and can be replaced by a constant $\mathbf{c}$. The exact computation of the constant can be challenging due to the integral involved. However, note that $\mathbf{c}$ is the normalizing constant for the posterior density, i.e., a constant that allows the density to integrate to 1. This information can sometimes be used to identify the constant $\mathbf{c}$. Now, recognizing $f(\mathbf{x}|\theta)$ as the likelihood function $L(\theta|\mathbf{x})$, we have

$$g(\theta|\mathbf{x}) = \mathbf{cL}(\theta|\mathbf{x})\pi(\theta) \tag{1.27}$$

Therefore, the posterior density of θ given the data $\mathbf{x}$ is an update of the likelihood of the data based on the prior density of θ.

To illustrate the idea, consider the vaccination trial example discussed in the earlier section, where θ is the true success rate of a vaccine. Consider a $Beta(\alpha, \beta)$ prior density of θ, i.e.,

$$\pi(\theta) = \frac{\Gamma(\alpha+\beta)}{\Gamma(\alpha)\Gamma(\beta)}\theta^{\alpha-1}(1-\theta)^{\beta-1},\ \theta > 0$$

Let x be the observed number of successes, we have the posterior

$$g(\theta|x) \propto \underbrace{\theta^{\alpha-1}(1-\theta)^{\beta-1}}_{Prior} \times \underbrace{\theta^{x}(1-\theta)^{n-x}}_{Binomial Likelihood} \propto \theta^{x+\alpha-1}\theta^{n-x+\beta-1} \tag{1.28}$$

It is clear that the posterior is given by a $Beta(x+\alpha, n-x+\beta)$ density with the proportionality constant c, to ensure that the density integrates to 1, is given by

$$c = \frac{\Gamma(n+\alpha+\beta)}{\Gamma(x+\alpha)\Gamma(n-x+\beta)}$$

The choice of prior density is an important computational and conceptual question in Bayesian inference, but a detailed exposition of this topic is beyond the scope of this discussion. However, a common choice of prior density is the *conjugate prior*. A prior is considered conjugate to a likelihood if the posterior belongs to the same family of probability density as the prior. It should be clear from the Binomial example discussed earlier, the Beta prior is conjugate to the Binomial family. The choice of conjugate prior simplifies the computational burden associated with Bayesian inference. Also, as shown in Figure (1.3), various choices of Beta parameters can be used to convey different prior information about the parameter θ. For example, the $Beta(1,1)$, which is the same as the $Uniform(0,1)$ distribution, conveys a lack of specific knowledge about the parameter, and in that sense, it is a non-informative choice. On the other hand, the $Beta(4,2)$ prior, which is a skewed-left distribution with a mean of 0.67, reflects the researcher's subjective belief that a larger value of θ is more likely. Finally, it is worth noting that $Beta(0.5,0.5)$ is a form of non-informative prior, called Jeffreys prior, which is invariant under the reparametrization of the model.

FIGURE 1.3
Various Beta priors.

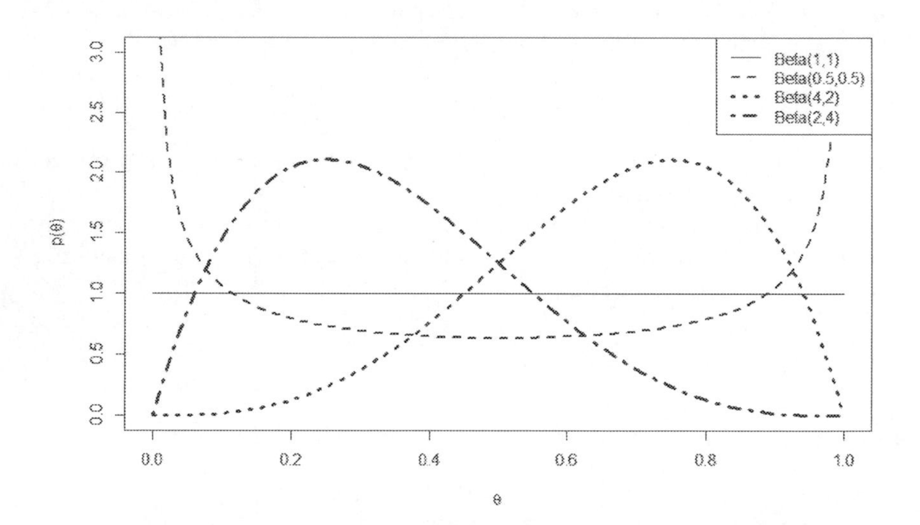

The posteriors distribution plays a central role in Bayesian inference. For example, A $(1-\alpha)$ Bayesian confidence interval $C(x)$, called a *Credible Interval*, for a parameter θ is given by

$$P\left(\theta \in C(x)|X=x\right)=1-\alpha. \tag{1.29}$$

In other words, the interval (or the region) $C(x)$ contains $(1-\alpha)$ posterior probability. It is useful to note the distinction between a frequentist confidence interval and a Bayesian credible interval. The *confidence* of a confidence interval is not about a specific interval obtained from a sample data; rather, it reflects confidence in the process of generating the interval in the sense that in the long run, the proportion of coverage of the true population parameter is at least $(1-\alpha)$. By contrast, a 95% credible interval states that the (posterior) probability that the parameter θ belongs to the credible interval is 0.95. Therefore, a credible interval has a simpler interpretation than a frequentist confidence interval.

The straightforward approach to the construction of a $(1-\alpha)$ credible interval from a posterior $g(\theta|x)$ is the equal-tail interval. Let g_α be the $(1-\alpha)$ percentile of the posterior distribution, i.e.,

$$P(\theta \leq g_\alpha | X = x) = 1 - \alpha \tag{1.30}$$

Then, an equal-tailed $(1-\alpha)$ credible interval is given by $(g_{\alpha/2}, g_{1-\alpha/2})$.

Consider the vaccine trial example discussed in the earlier sections, where θ is the success rate in eliciting antibodies in the subjects at a certain level. Let the number of participants in the study is $n = 20$, out of which vaccination was successful in $x = 18$ subjects. Consider the non-informative Jeffreys prior $Beta(0.5, 0.5)$. Therefore, from (1.28), the posterior is given by the $Beta(18.5, 2.5)$ distribution, which is displayed in Figure (1.4), and the corresponding 95% equal-tailed credibility interval is given by $(0.7161, 0.9786)$.

FIGURE 1.4
Beta(18.5,2.5) posterior and the 95% equal-tail credible interval.

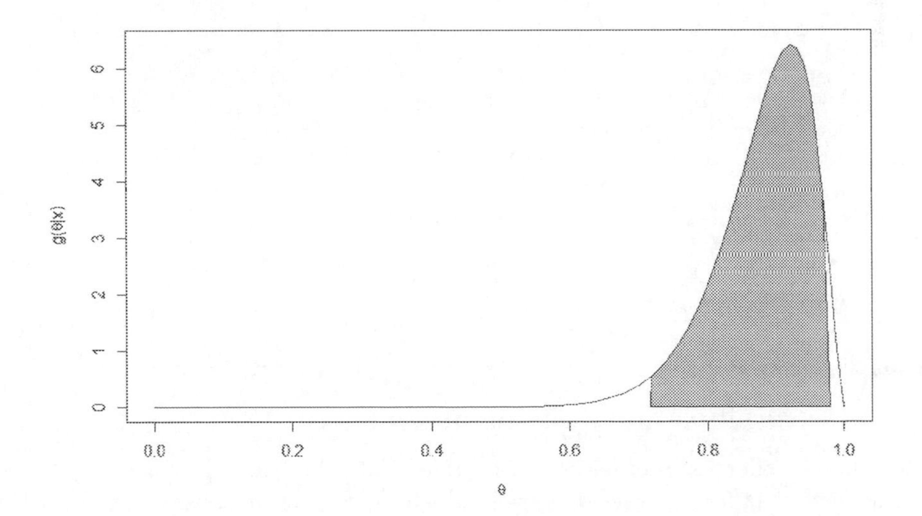

The equal-tail credible internal is not the shortest interval unless the posterior distribution is symmetric. In this example, where the posterior is skewed, the *Highest Posterior Density* or *HPD* interval is preferable. Suppose $C(x)$ is an HPD confidence interval (or region). In that case, the posterior density for every point in this interval is higher than the posterior density for any point outside of this interval. Figure (1.5) illustrates an HPD interval for a unimodal posterior distribution. It should be intuitively clear that the $(1-\alpha)$ HPD interval is the shortest $(1-\alpha)$ credible interval. It is also worth noting that a credible interval is not necessarily an interval. If the posterior distribution is multimodal, the HPD region is given by the union of intervals. However, in this book, we will loosely use the term interval in describing a credible region. In the vaccine study described in the last paragraph with a posterior distribution $Beta(18.5, 2.5)$, the 95% HPD interval is given by $(0.7460, 0.9907)$, which is narrower than the equal-tail interval, as expected.

FIGURE 1.5
Beta(18.5,2.5) posterior and the 95% HPD interval.

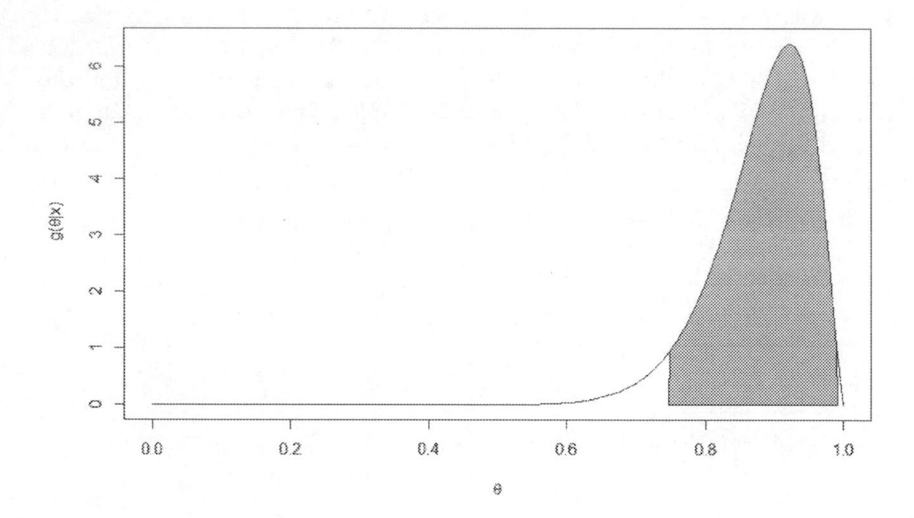

1.4 Discussions and conclusions

This chapter offers a brief overview of inferential procedures in statistics. It is expected that a reader of this book will have a general understanding of statistical inference and modeling methods, and this chapter is meant to reacquaint the readers with the concepts and terminology utilized in the chapters that follow.

The book focuses on the analysis of discrete data. This is a critical area of inferential statistics but especially significant in the context of clinical research. In this regard, there is no shortage of methodologies to answer specific questions, which is a blessing and a curse. The choices available to a researcher are always welcome, but the lack of available information about the comparative merits of these methods certainly hinders the decision-making process of the investigator. The book offers a menu of available methods in the literature to address specific inference questions and provides some clarity in terms of the relative utility of the methods.

2

Are We Slaves to the p-Value: The ASA's Statement on p-Value

2.1 Introduction

Of all the statistical methods commonly used, hypothesis testing, and the related concept of "statistical significance", is perhaps the most important because of its widespread use in all branches of science for drawing inference. It has undoubtedly a far-reaching implication in shaping up the future of science and, more generally, the future of human society. In spite of its widespread use, p-value is often misinterpreted and misused for drawing scientific inference. The American Statistical Association (**ASA**), therefore, issued a much-required and timely statement in 2016 [97] on the appropriate use of p-values for scientific inference. The statement triggered a whole lot of insightful discussions around it by the experts, which is published as supplemental materials. This chapter summarizes the key principles laid out in the **ASA** statement and a few important issues that came out of the discussion.

The **ASA** statement defines the p-value as, "*the probability under a specified statistical model that a statistical summary of the data (e.g., the sample mean difference between two compared groups) would be equal to or more extreme than its observed value.*" This means, the smaller the p-value, the higher is the significance, and more incompatibility between the specified model (often called the null hypothesis) and the data. The "usual scientific practice" is to reject the null hypothesis if the p-value is less than or equal to a certain pre-defined, but arbitrary threshold, which is referred to as the level of significance. In practice, the most commonly used values of the level of significance are 0.05, 0.01, 0.005 and 0.001. Note that the threshold is not based on any analysis of the observed data, any scientific or quantitative rationale, or the underlying hypothesis in question. It is often a widely accepted number for similar problems.

The dichotomous decision, "reject" or "do not reject" based on p-value, is the root cause of all its misuse and misinterpretation. Fisher proposed p-value as a measure of evidence against the null hypothesis, the lower the p-value, the greater is the likelihood that the null hypothesis is false, and thus, stronger

DOI: 10.1201/9781315169859-2

is the evidence against the null hypothesis. On the other hand, the Neyman–Pearson approach to the null hypothesis testing (**NHT**) introduced testing as a two-decision problem, either "reject" or "do not reject" the null hypothesis. Fisher, indeed, was highly critical of this approach. However, without a thorough understanding of either approach, the non-statisticians began to write statistics manuals for working scientists, and created a "hybrid system that crammed Fisher's easy to calculate p-value into Neyman–Pearson's reassuringly rigorous rule-based system" (Nuzzo [65]). Unfortunately, till today, this hybrid system is used by almost all the scientists in almost all the branches of science for drawing scientific inference.

2.2 ASA statement on statistical significance and p-values

The **ASA** statement lays out a few agreed upon principles for ensuring proper use and interpretation of p-values in drawing scientific reporting. The principles provide key guidance on the use of p-values, and have been further discussed in the accompanying commentaries.

1. p-values can indicate how incompatible the data are with a specified statistical model

It states that the p-value is only indicative of the incompatibility between the data and the specified model. From the definition of p-value, it is obvious that smaller the p-value, more is the statistical incompatibility of the data with the null hypothesis, of course, if the underlying assumptions used to calculate the p-value hold.

2. P-values do not measure the probability that the studied hypothesis is true, or the probability that the data were produced by random chance alone

Often, the researchers misinterpret p-value as the probability of the truth of a null hypothesis, or the observed data being produced by random chance alone. Both are wrong. "It is a statement about data in relation to a specified hypothetical explanation, and is not a statement about the explanation itself" ([97]). As Stark [86] clarifies, "A p-value is computed by assuming that the explanation is right. The p-value is *not* the probability that the explanation is right. For example, if our hypothesis tests the treatment effect, the p-value is the probability of obtaining an effect at least as extreme as the one in the collected sample data, assuming the truth of the null hypothesis. It is not the probability of the assumption itself being true".

3. Scientific conclusions and business or policy decisions should not be based only on whether a p-value passes a specific threshold

The cause of p-value misinterpretation is the "p-value fallacy", a term coined by Steven N. Goodman in 1999 [43]. The fallacy arises from using a mechanical "bright line" rules (such as "reject" if p-value ≤ 0.05, "do not reject", otherwise) for justifying scientific claims or conclusions. A conclusion cannot be "true" on the one side of the line, and "false" on the other. For drawing conclusion, a researcher should take into account various other important contextual factors that may have a bearing on the scientific inference, like "the design of a study, the quality of the measurements, the external evidence for the phenomenon under study, and the validity of assumptions that underlie the data analysis" ([97]). Therefore, conclusions drawn based only on the "bright line" rule without taking into consideration other supporting evidence are heavily criticized. This can not only lead to erroneous conclusions but also "degrades vast efforts to collect and analyze quantitative data into a mere label" (Rothman [77]).

In this regard, a widely agreed upon principle is to consider p-value as an evidential measure in continuum rather than as an input to a binary "yes-no" decision. "Pragmatic considerations often require a binary, 'yes-no' decisions, but this does not mean that p-values alone can ensure that a decision is correct or incorrect" ([97]). For ensuring proper scientific inference, this is, perhaps, the most crucial principle in the **ASA** statement.

4. Proper inference requires full reporting and transparency

A clarification of the principle follows ([97]), "P-values and related analyses should not be reported selectively. Conducting multiple analyses of the data and reporting only those with certain p-values (typically those passing a significance threshold) renders the reported p-values essentially uninterpretable. Cherry-picking promising findings, also known by such terms as data dredging, significance chasing, significance questing, selective inference, and 'p-hacking' leads to a spurious excess of statistically significant results in the published literature and should be vigorously avoided".
The authors of the **ASA** statement further add "... whenever a researcher chooses what to present based on statistical results, valid interpretation of those results is severely compromised if the reader is not informed of the choice and its basis. Researcher should disclose the number of hypotheses explored during the study, all data collection decisions, all statistical analyses conducted, all p-values computed. Valid scientific conclusions based on

p-values and related statistics cannot be drawn without at least knowing how many and which analyses were conducted, and how those analyses (including p-values) were selected for reporting" ([97]).

In the context of drug development, Berry [7] has several important comments to make. "P-values ignore many aspects of the evidence in the experiment at hand including information that is obviously known... The specifics of data collection and curation and even your intentions and motivation are critical for inference. What have you not told the statistician? Have you deleted some data points or experimental units, possibly because they seemed to be outliers? Are some entries actually the average of two or more measurements made on the same experimental unit? If so, why were there more measurements on some units than others? Have you conducted other experiments addressing the same or related questions and decided that this was the most relevant experiment to present to the statistician? And on and on and on. The answers to these questions may be more important for making inferences than the numbers themselves. They set the context for properly interpreting the numerical aspects of the 'data'. Viewed alone, p-values calculated from a set of numbers and assuming a statistical model are of limited value and frequently are meaningless".

Berry [7] further suggests, "When there is a prospective study protocol and statistical analysis plan then both should be made available at the time of publication with any deviations from the original plan. In the absence of a protocol and statistical analysis plan the credibility of conclusions is low, despite honest attempts to say what analyses had been planned, whether done or not, and what planned analyses were not done. And adjusting for the associated multiplicities may be difficult in this circumstance. A pragmatic approach is to completely describe the multiplicities, keeping a log of what was done, and then giving 'unadjusted' p-values, including a black-box warning, 'Our study is exploratory and we make no claims for generalizability. Statistical calculations such as p-values and confidence intervals are descriptive only and have no inferential content"'.

Finally, it's worth mentioning that the present editorial policy of most of the scientific journals, incentivizes researchers to "Cherry-picking promising findings." Most of the journals tend to publish the research with statistically significant (positive) results, and that too using the mechanical "bright line" rules as mentioned in Principle 3. A change in the editorial policy is essential for discouraging researchers from "Cherry-picking promising findings" and, thus, help in bringing a change in the way the scientific inference is made now.

5. A p-value, or statistical significance, does not measure the size of an effect or the importance of a result

The **ASA** statement ([97]) further clarifies, "Statistical significance is not equivalent to scientific, human, or economic significance. Smaller p-values do not necessarily imply the presence of larger or more important effects, and larger p-values do not imply a lack of importance or even a lack of effect. Any effect, no matter how tiny, can produce a small p-value if the sample size or measurement precision is high enough, and large effects may produce unimpressive p-values if the sample size is small or measurements are imprecise. Similarly identical estimated effects will have different p-values if the precision of the estimates differ".

To explain the principle, let's consider a simple testing problem. Suppose, based on a sample of size n from a normal population with mean μ and known variance σ^2, the null hypothesis $\mu = \mu_0$ is to be tested against the alternatives $\mu > \mu_0$, where μ_0 is pre-specified. The standard rule is, reject the null hypothesis at level 0.05, if $Z = (\bar{x} - \mu_0)(\sigma/\sqrt{n})^{-1}$ is greater than 1.65, where $\bar{x}$ is the sample mean. If the observed value of Z is 1.65, the p-value is approximately 0.05, and if Z is 2.34, the p-value is 0.01. In general, the p-value decreases as observed value of Z increases. The observed value of Z is the product of the estimated effect size $(\bar{x} - \mu_0)/\sigma$ and $\sqrt{n}$. Thus for a very small estimated effect size, Z could be made very large, and thus, the resulting p-value very small, by making the sample size n sufficiently large. On the other hand, for a much larger effect size, the p-value could be considerably larger if the sample size is small. What is important to note here is that the p-value heavily depends on the sample size, but effect size $(\mu - \mu_0)/\sigma$ does not. Thus, making scientific inferences solely based on p-value may lead to a poor decision.

Sullivan and Feinn ([88]) provides an illuminating discussion on this issue with elementary examples, and is specially useful for readers with little statistical background. It recommends the researchers to use p-value and effect size together to describe their findings.

6. By itself, a p-value does not provide a good measure of evidence regarding a model or hypothesis

Principle 6 is a kind of caveat to the researchers about the limitation of p-values as a sole measure of evidence for scientific inference. As a clarification, the **ASA** statement ([97]) adds, "Researchers should recognize that a p-value without context or other evidence provides limited information. .... For these reasons, data analysis should not end with the calculation of a p-value when other approaches are appropriate and feasible."

In this context, many experts suggest that, in addition to sharing all the relevant information about the collection, measurements and analyses of data, the p-values should be complemented with confidence intervals and effect-size estimates for making a better scientific inference (Benjamini ([6])). The American Statistician supplemental material on the ASA statement adds, "The p-value is a valuable tool, but when possible it should be complemented-not replaced-by confidence intervals and effect size estimates. The end of a 95% confidence interval that extends towards 0 indicates by how much the difference can be separated from 0 (in statistically significant way at level 5%...)". Millar [58] notes, "At a minimum we can include confidence intervals whenever we perform a test, to assess practical significance in addition to statistical significance".

2.3 Discussion and recommendation

In this chapter, we summarize the recommendations made by the **ASA** on statistical significance and p-values, and also discuss some of the issues arising out of it. The **ASA** stepped in, and issued the statement as a reaction to the recent attacks on p-values and the role of statistical significance in the "crisis of irreproducibility" ([28]). Berry ([8]) describes the situation nicely: "Our collective credibility in the science community is at risk. We cannot excuse ourselves by blaming non-statisticians for their failure to understand or heed what we tell them. ... We must communicate better even if we have to scream from the rooftops, which is exactly what the **ASA** is doing".
Goodman ([44]), in his commentary, raises perhaps, the most important issue. It is the last mile, the challenge of maximizing the adoption of these principles by the scientific community on the ground. "Almost none of the (**ASA**) statement is new .... some of these same principles were stated or argued about at least a century ago, with many reminders between then and now. So the question we must ask ourselves is, how could this have happened, and what can we do to change it? Write new textbooks? New software? Overhaul the teaching of statistics? Change the journal review policy? Change the regulatory policies for drug approval? Many think not much is going to happen, at least in near future."

In the present context, let us be explicit about what this book can offer to the readers. Given the **ASA** recommendations, in the subsequent chapters, are we still going to follow the traditional approach of drawing inference based on the p-values? Let's quote Gelman's (Supplementary materials) thoughts on this issue, "it seems to me that statistics is often sold as a sort of alchemy that transmutes randomness into certainty, an 'uncertainty laundering' that begins with data and concludes with success as measured by statistical significance. ... We present neatly packaged analyses with clear conclusions. This is what is expected - demanded - of subject matter journal". In this book also, we adopt the traditional approach of statistical significance based on the p-values. For us, the reasons are all the more strong. As stated in the preface, first, this is a cookbook of discrete data analysis primarily written for the practicing statisticians in pharmaceutical industries, who wish to learn and implement the toolbox of available confidence intervals for significance testing. Most importantly, the regulatory authorities, like FDA, have still been practicing the "reject-do not reject" binary rules for statistical significance based on p-values. Having said so, at the same time, we would like to make the readers aware of the **ASA** recommendations, and appreciate the limitations of the p-values for making scientific inference. We sincerely hope that the discussion presented in this chapter will motivate the readers to think differently.

3

One Binomial Proportion

3.1 Introduction

In this chapter, we consider the methodologically important problem of estimation of a binomial proportion. This is a basic yet challenging inferential question that requires careful consideration. The commonly and often uncritically used confidence interval for a binomial proportion p, also known as Wald's confidence interval, is given by $\widehat{p} \pm z_{\alpha/2}\sqrt{\widehat{p}(1-\widehat{p})/n}$, where $\widehat{p}$ is the sample proportion, n denotes the sample size and $z_{\alpha/2}$ is the $(1-\frac{\alpha}{2})$ quantile of a standard normal distribution. It is well established in the literature that the coverage properties of the interval are erratic, and the ad-hoc remedies advocated in textbooks, such as $n\widehat{p} \geq 5$, and $n(1-\widehat{p}) \geq 5$, does not mitigate the problem.

In this chapter, we will discuss wide-ranging methods of computing confidence intervals of a single binomial proportion. The methodologies discussed in this chapter can be broadly classified into three groups, namely, asymptotic, Bayesian and exact methods. The first part of the chapter discusses five different asymptotic methods: Wald, Wald with continuity correction (AS) (Fleiss et al. [36]), Wilson (SC) [99], Wilson with continuity correction (SCC) (Fleiss et al. [36]), Agresti-Coull (AC) [3] and second-order corrected (SOC) one-sided CI (Cai [17]). The next part of the chapter discusses Bayesian intervals using Jeffreys' prior (JF) (Brown et al. [15]) before discussing four exact interval methods, namely, Clopper–Pearson (CP) [26], mid-p adjusted Clopper–Pearson (MCP) [4], Blyth–Still–Casella (BS) (Blyth and Still [13], Casella [20], and Blaker (BL) [11]. The performances of these intervals are discussed in the subsequent section in light of two criteria, viz., the nominal coverage probability and expected length of the intervals. Finally, the chapter concludes with some discussions and recommendations.

3.2 Testing of a hypothesis

Let a proportion p_0 be the reference value or the *null* value to be tested for a proportion p. Suppose x is the observed number of success in n independent

DOI: 10.1201/9781315169859-3

Bernoulli trials, then the true proportion p can be estimated as $\widehat{p} = \frac{x}{n}$. Let's consider the following two-sided and one-sided hypotheses:

1. The two-sided test – test of equality

$$H_0 : p = p_0 \quad \text{versus} \quad H_1 : p \neq p_0 \tag{3.1}$$

   Let $\epsilon = p - p_0$, then above can be re-written as the following:

$$H_0 : \epsilon = 0 \quad \text{versus} \quad H_1 : \epsilon \neq 0 \tag{3.2}$$

2. The one-sided upper-tailed test:

$$H_0 : p \leq p_0 \quad \text{versus} \quad H_1 : p > p_0 \tag{3.3}$$

3. The one-sided lower-tailed test:

$$H_0 : p \geq p_0 \quad \text{versus} \quad H_1 : p < p_0 \tag{3.4}$$

In applied research, mainly in the clinical trial application, the one-sided tests are frequently used in testing **non-inferiority** or **superiority** of a reference proportion p_0. In such a setting, suppose δ is the margin for non-inferiority or superiority. Then for testing the non-inferiority or the superiority, the following unified hypothesis can be tested:

$$H_0 : p - p_0 \leq \delta \quad \text{versus} \quad H_1 : p - p_0 > \delta \tag{3.5}$$

or

$$H_0 : \epsilon \leq \delta \quad \text{versus} \quad H_1 : \epsilon > \delta \tag{3.6}$$

When $\delta > 0$, the hypothesis to be tested is called the superiority test over the reference value p_0 (by margin δ), and similarly when $\delta < 0$, the hypothesis to be tested is called the non-inferiority test over the reference value p_0 (by margin δ) (see Chow, Shao and Wang [25] for further details).

An equivalence test is used to demonstrate the similarities or lack of differences against a reference value of p_0. For this, two one-sided hypotheses are used as given in the following (see Chow, Shao and Wang [25] for further details).

$$H_0 : |p - p_0| \geq \delta \quad \text{versus} \quad H_1 : |p - p_0| < \delta \tag{3.7}$$

or

$$H_0 : |\epsilon| \geq \delta \quad \text{versus} \quad H_1 : |\epsilon| < \delta \tag{3.8}$$

3.3 Asymptotic confidence intervals

Suppose X is the number of successes in n independent Bernoulli trials with probability of success p. Let x be the observed value of X, and we denote the observed success rate $\frac{x}{n}$ by $\widehat{p}$. Furthermore, let a two-sided $100 \times (1-\alpha)\%$ confidence interval of p is given by $(\underline{\Delta}, \overline{\Delta})$. Then, a one-sided $100(1-\alpha/2)\%$ CI is obtained from the two-sided CI by considering either $(max(0, \underline{\Delta}(X)), 1)$ or $(0, min(1, \overline{\Delta}(X)))$.

3.3.1 Wald confidence interval

The universally accepted method of CI for p included in almost all standard statistics textbooks is the well-known Wald's confidence interval of a binomial proportion. For an estimate $\widehat{p} = \frac{x}{n}$, the two-sided $100(1-\alpha)\%$ confidence interval is computed as follows:

$$\widehat{p} \pm z_{\alpha/2}\sqrt{\widehat{p}(1-\widehat{p})/n}$$

The confidence interval is derived from the Central Limit Theorem (CLT). If the "conditions of CLT hold", then $(\widehat{p}-p)/\sqrt{\widehat{p}(1-\widehat{p})/n}$ approximately follows a standard normal distribution. Therefore, the probability of the event $-z_{\alpha/2} \leq (\widehat{p}-p)/\sqrt{\widehat{p}(1-\widehat{p})/n} \leq z_{\alpha/2}$ is roughly $(1-\alpha)$. Wald's CI for p can be easily derived by inverting this inequality. However, the method's success depends on n and p, and a range of different conditions have been proposed to ensure that the normal approximation holds. These conditions include recipes, such as (i) the interval $\widehat{p} \pm 3\sqrt{\widehat{p}(1-\widehat{p})/n}$ does not include 0 or 1, (ii) $n\widehat{p} \geq 15 \text{should be non} - \text{italic}, n(1-\widehat{p}) \geq 15$, (iii) n "large", among many others. These conditions, unfortunately, do not address the shortcomings of Wald's method.

As discussed in chapter 1, a common metric used to assess the performance of a method for computing a confidence interval is the coverage probability. This term refers to the probability that a procedure for constructing interval estimates will produce an interval containing, or covering, the true value of the population parameter. Brown et al. [14] discussed a peculiar erratic behavior of the coverage probability of Wald's method, which they termed as the "lucky-n lucky-p" phenomenon. In essence, for a given p, the coverage probability of Wald's confidence interval is not monotone increasing in n and depends on specific choices of n and p. Figure 3.1 shows the effect of different choices of p on the coverage probability of a nominal 95% confidence interval for $n = 20$. The plot exhibits an oscillating behavior in p, and some "lucky" p, such as $p = 0.144$, yields a coverage probability of approximately 0.95, whereas the coverage probability for an "unlucky" $p = 0.16$ is about 0.84. The oddity of the situation is that a larger sample size does not guarantee a better coverage probability. Figure 3.2 shows the effect of sample size on the

FIGURE 3.1
Coverage probabilities of Wald confidence intervals.

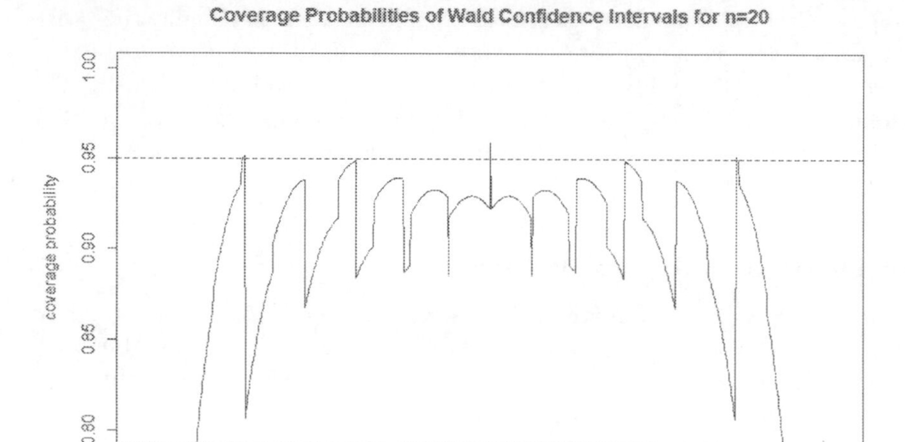

coverage probability of Wald CI for $p = 0.5$, for which one would expect the CI to attain the nominal confidence level even for a moderate sample size. However, the plot again shows an erratic behavior of the coverage probability for sample sizes ranging from 10 to 100. The coverage probability for $n = 17$ actually attains the nominal probability of 0.95 but surprisingly drops down to 0.93 for a much larger sample size $n = 52$.

This clearly demonstrates the limitations of the popular Wald confidence interval. The lack of reliability of Wald CI in terms of its coverage probability is disconcerting, to say the least, and makes the method unsuitable for critical applications in clinical research.

Illustration: Asymptotic Wald intervals

Let's consider an example from a medical device trial (Pradhan et al. [69]). In this trial, 31 women patients were tested with a medical device to treat their stress urinary continence (SUI), a disease where women were unable to hold their urine, resulting in leaks due to minor physical activities like coughing, walking, etc. In the trial, 28 patients reported success and 3 reported failure after 12 months. Based on the historical data, the SUI success rate was assumed to be 85% with other methods. The study was intended to test the non-inferiority of the device with a 10% margin with $\alpha = 0.025$. Hence the statistical hypothesis was set to be tested as follows:

FIGURE 3.2
Coverage probabilities of Wald confidence intervals for $p = 0.5$.

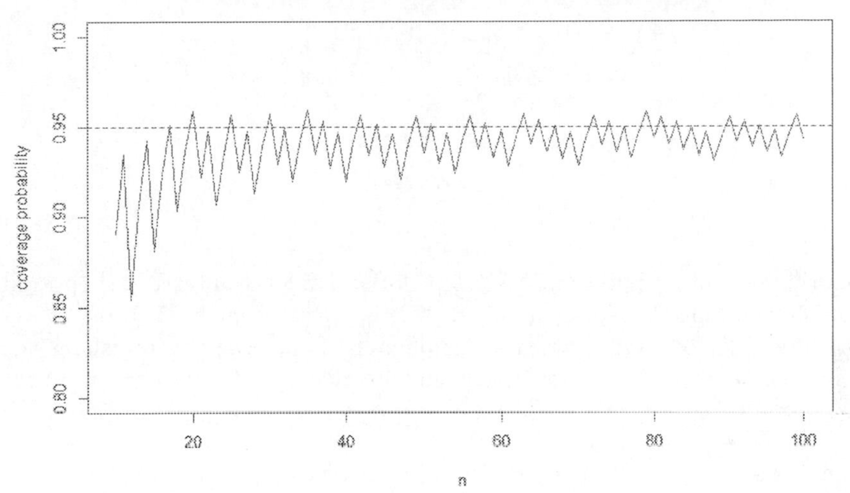

$$H_0 : p \leq 0.75 \quad \text{versus} \quad H_1 : p > 0.75$$

In testing non-inferiority or superiority, often confidence interval approaches are recommended by regulatory agencies. Specifically, in this example, for the one-sided test being considered, the lower limit of the 95% interval is compared against 0.75. If the lower limit of the 95% interval is greater than 0.75, then the null hypothesis is rejected at 5% level of significance.

Almost all statistical software packages can compute the Wald confidence interval. Since the regulatory submissions require outputs using SAS software, this chapter mainly focuses on the results obtained from SAS. The Wald confidence interval can be computed in `SAS` using `PROC FREQ` and the following code:

```
ods listing;
data sui;
input trt response freq;
cards;
1 1 28
1 2 3
;
run;
proc freq data=sui ;
tables response/nocum norow binomial (cl=wald) alpha=0.05;
weight freq;
run;
```

In the SAS code, the option *alpha=0.05* is for computing a 95% confidence interval. It should be noted that by default SAS computes 95% confidence interval, hence one can omit this option. The specification *binomial (cl=wald)* ensures that Wald's method is used for the confidence interval calculation. The SAS output is given below:

```
            Binomial Proportion
               response = 1

Proportion                   0.9032
ASE                          0.0531
95% Lower Conf Limit         0.7992
95% Upper Conf Limit         1.0000
```

In this partial output from SAS, the 95% confidence interval is reported as $(0.7992, 1)$. Since the lower limit, 0.7992, is greater than 0.75, the null hypothesis is rejected, and statistical significance is achieved establishing the non-inferiority of the device in testing the effectiveness of the device in treating SUI.

In chapter 1, we have discussed the duality between confidence interval and hypothesis testing procedures, i.e., one can decide whether or not to reject the null hypothesis based on the corresponding confidence interval. However, if the objective is to compute the p-value of the hypothesis test, then one can use the following SAS code:

```
ods listing;
data sui;
input trt response freq;
cards;
1 1 28
1 2 3
;
run;
proc freq data=sui ;
tables response/nocum norow binomial (cl=wald p=0.85 noninf margin=0.1) alpha=0.025;
weight freq;
run;
```

In the above SAS PROC FREQ code *(cl=wald p=0.85 noninf margin=0.1) alpha=0.025* specifies non-inferiority $margin = 0.1$ for testing $H_0 : p = 0.85$ with $\alpha = 0.025$. The SAS output shows the following:

```
Confidence Limits for the Binomial Proportion
            Proportion = 0.9032

      Type    97.5% Confidence Limits

      Wald    0.7842          1.0000


          Noninferiority Analysis

H0: P - p0 <= -Margin    Ha: P - p0 > -Margin
```

```
                    p0 = 0.85   Margin = 0.1

          Proportion  ASE (Sample)         Z    Pr > Z

              0.9032        0.0531    2.8856    0.0020
```

The *Noninferiority Analysis* section of the output states the null and alternative hypothesis, and the corresponding p-value is reported as 0.002, which is highly significant. Note that when the Wald method is used, the inference using p-value will match with the corresponding confidence interval, as the confidence interval is constructed by inverting the Wald test. However, as discussed earlier, due to poor coverage probability, Wald interval is not recommended for sensitive clinical applications, particularly when the sample size is small (Vollset [95]). Since the sample size $n = 31$, in this context, is considered a small sample, the inference based on the Wald method may not be valid. Additional details on the coverage and expected lengths are given in the latter part of this chapter.

3.3.2 Continuity corrected Wald interval

This continuity corrected CI given in Fleiss et al. [36] is obtained by modifying the test statistic for the Wald test. Let $z_{\frac{\alpha}{2}}$ denote the $100(1-\alpha/2)$-th percentile of the standard normal distribution, then the continuity corrected CI is given by:

$$\widehat{p} \pm z_{\frac{\alpha}{2}} \sqrt{\widehat{p}(1-\widehat{p})/n} + \frac{1}{2n}$$

The coverage of continuity-corrected Wald interval is shown in Figure 3.3.

The coverage plot shows that the continuity correction to the Wald method did not improve the coverage significantly. The continuity-corrected Wald interval (AS) has poor coverage property with a small sample and is strongly discouraged for critical applications (Vollset [95]). However, we have discussed the procedure here for the sake of completeness, as most standard statistical packages (for example, `SASs PROC FREQ`) implement the method.

Illustration: Continuity corrected Wald interval

Consider the SUI example introduced in Section 3.3. Using SAS PROC FREQ, the Wald with continuity correction method can be employed using the following SAS code:

```
ods listing;
data sui;
input trt response freq;
cards;
1 1 28
1 2 3
;
run;
```

FIGURE 3.3
Coverage probabilities of Wald and Wald (corrected) confidence intervals for $n = 20$.

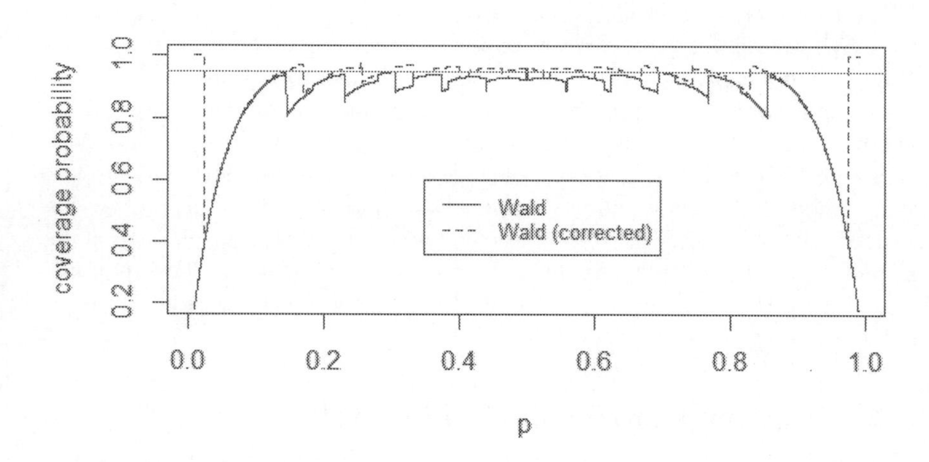


```
proc freq data=sui ;
tables response/nocum norow binomialc (cl=wald) alpha=0.05;
weight freq;
run;
```

In the code, the SAS keyword *binomialc* is for continuity correction method for the *cl=wald* option. When the code is executed in SAS, the partial output include the following:

```
            Confidence Limits for the Binomial Proportion
                         Proportion = 0.9032

             Type                 95% Confidence Limits

             Wald (Corrected)     0.7830          1.0000
```

In the output, the Wald method with continuity correction shows the 95% confidence interval as $(0.7830, 1.0000)$. Again, for hypothesis testing, since the lower limit 0.7830 is greater than 0.75, the null hypothesis is rejected, and the non-inferiority of the device efficacy is established.

3.3.3 Wilson score interval

Based on the score statistic, Edwin B. Wilson [100] first proposed this method in 1927, and this is one of the recommended asymptotic methods of computing the confidence interval of a single binomial proportion in the literature

(Newcombe [63], Brown Cai and DasGupta[14]). The interval is constructed by equating the score statistic z to its critical values $z_{\frac{\alpha}{2}}$, i.e.,

$$\frac{\widehat{p} - p}{[\frac{p(1-p)}{n}]^{\frac{1}{2}}} = \pm z_{\frac{\alpha}{2}}.$$

The confidence interval due to Wilson (also known as score interval) is a solution to the quadratic equation in p, and the confidence limits are computed as follows:

$$w\widehat{p} + \frac{1}{2}(1-w) \pm z_{\frac{\alpha}{2}} \left[w^2 \frac{\widehat{p}(1-\widehat{p})}{n} + (1-w)^2 \frac{(1/2)(1/2)}{z^2_{\frac{\alpha}{2}}} \right]^{\frac{1}{2}},$$

where $w = n/(n + z^2_{\frac{\alpha}{2}})$.

Even with smaller sample sizes, the confidence intervals due to Wilson are known to have good coverage probabilities with shorter expected lengths. Figure 3.4 shows the coverage probabilities of the Wilson CI for $n = 20$. Although the oscillating feature observed in the Wald interval is still present, it is clear that for most p, the coverage probabilities are greater than the nominal 95%, and a "lucky" p is much more likely in the Wilson CI compared to the Wald CI.

FIGURE 3.4
Coverage probabilities of Score (Wilson) confidence intervals for $n = 20$.

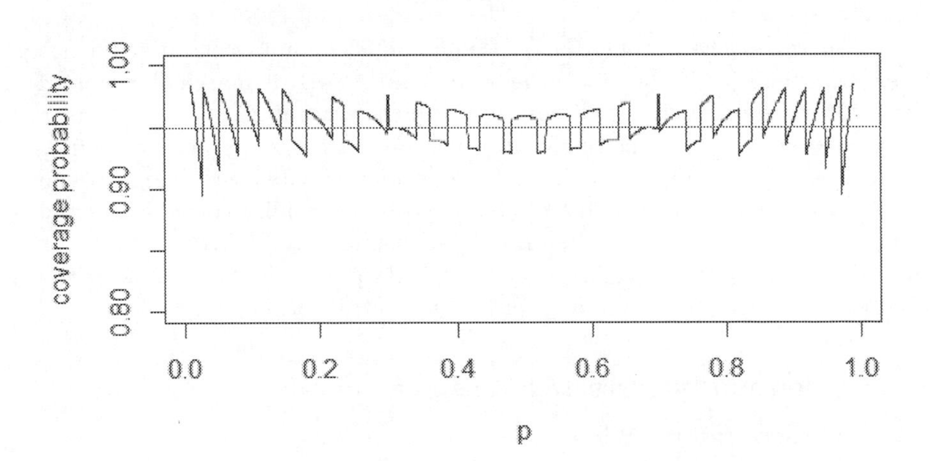

Illustration: Wilson score interval

To illustrate Wilson's method, consider the SUI example introduced in Section 3.3. The method can be implemented in SAS PROC FREQ, as follows:

```
ods listing;
data sui;
input trt response freq;
cards;
1 1 28
1 2 3
;
run;
proc freq data=sui ;
tables response/nocum norow binomial (cl=wilson) alpha=0.05;
weight freq;
run;
```

The SAS keyword *binomial (cl=wilson)* invokes the one-way binomial confidence interval method due to Wilson (1927). The option *cl=wilson* included in the code for Wilson's score interval; alternatively, one could also use *cl=score*. A partial SAS output is given below.

```
                    The FREQ Procedure

                   Binomial Proportion
                      response = 1

                  Proportion        0.9032
                  ASE               0.0531

        Confidence Limits for the Binomial Proportion
                   Proportion = 0.9032

              Type       95% Confidence Limits

              Wilson     0.7510           0.9665
```

As shown in the output, the Wilson method computes 95% confidence interval as $(0.751, 0.9665)$. Since the lower limit, 0.751, is marginally greater than 0.75, the null hypothesis is rejected, and the non-inferiority of the device is established. However, the small sample size and closeness of the lower limit of the confidence interval to the null value of 0.75 calls into question the conclusion's validity. A confidence interval using continuity correction of the same method, if it exists, is desirable in such a situation. It turns out that there is a continuity correction version of Wilson's method available in the literature, and we discuss the approach in the next section.

3.3.4 Continuity corrected Wilson interval

The continuity corrected Wilson interval is given by:

$$w\widehat{p^\star} + \frac{1}{2}(1-w) \pm z_{\alpha/2}\left[w^2\frac{\widehat{p^*}(1-\widehat{p^\star})}{n} + (1-w)^2\frac{(1/2)(1/2)}{z^2_{\frac{\alpha}{2}}}\right]^{\frac{1}{2}}$$

which is obtained by replacing $\widehat{p}$ in the Wilson score interval by $\widehat{p}^{\star} = \widehat{p}+1/(2n)$ for the upper limit and $\widehat{p^{*}} = p - 1/(2n)$ for the lower limit. The lower and the upper limits of the interval are set at 0 and 1 for $\widehat{p}$ at 0 and 1, respectively. The coverage probabilities of Wilson and continuity corrected Wilson method Wilson (corrected) are given in Figure 3.5.

FIGURE 3.5
Coverage probabilities of Wilson and Wilson (corrected) confidence intervals for $n = 20$.

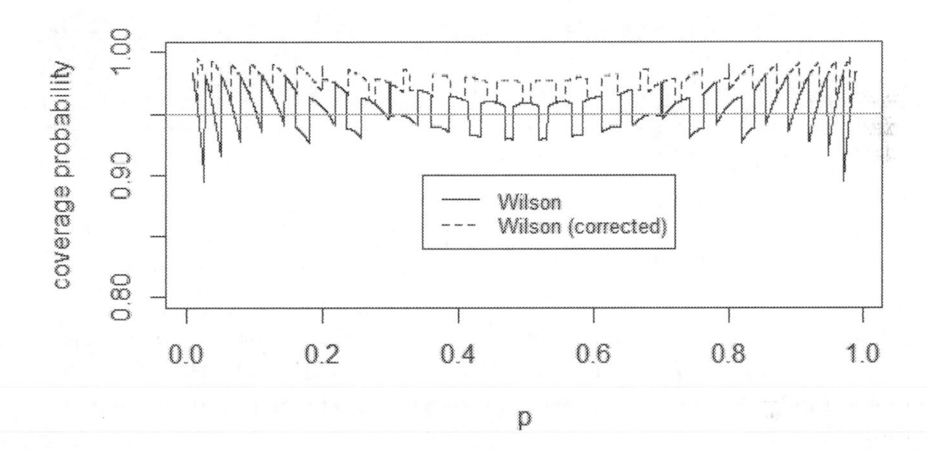

As seen in Figure 3.5, in general, the expected coverage probabilities using the Wilson method are good, mostly ranging around 95% level, but there are several instances of under coverages – mostly slightly below the 95% level. However, with the continuity correction method, most of these under coverages disappeared, as can be seen in Figure 3.5. In Figure 3.5, all the dotted lines (representing the coverage probabilities of Wilson (corrected) method) are well above the 95% level; this indicates that the Wilson with continuity correction method always guarantees the desired coverage level even for small samples.

Illustration: Asymptotic continuity corrected score interval due to Wilson (1927)

To illustrate the method, consider the SUI example introduced in Section 3.3. Using SAS PROC FREQ, the continuity corrected score interval due to Wilson (1927) can be employed using the following SAS code:

```
ods listing;
data sui;
```

```
input trt response freq;
cards;
1 1 28
1 2 3
;
run;
proc freq data=sui ;
tables response/nocum norow binomialc (cl=wilson) alpha=0.05;
weight freq;
run;
```

The SAS keyword *binomialc* is for continuity correction method for the option *cl=wilson*. A partial SAS output is given below.

```
                    Binomial Proportion
                        response = 1

                  Proportion        0.9032
                  ASE               0.0531

        Confidence Limits for the Binomial Proportion
                     Proportion = 0.9032

         Type                   95% Confidence Limits

         Wilson (Corrected)     0.7310          0.9747
```

The continuity corrected score interval due to Wilson shows 95% confidence interval as $(0.731, 0.9747)$. Since the lower limit 0.731 is less than 0.75, the null hypothesis cannot be rejected, and the non-inferiority of the device is not established.

3.3.5 Agresti and Coull interval

The Agresti and Coull [3] interval is obtained by adjusting the Wald interval. This is done by adding "two success and two failures" to the sample in the Wald interval when the nominal coverage probability is 0.95. At $\alpha = 0.05$, the adjusted point estimate of p is given by $\widehat{p}_{adj} = \frac{x+2}{x+4}$, which is approximately $(x + \frac{1}{2}z^2_{\alpha/2})/(n + z^2_{\alpha/2})$. Therefore, the resulting confidence interval is:

$$\widehat{p}_{adj} \pm z_{\alpha/2}\sqrt{\widehat{p}_{adj}(1 - \widehat{p}_{adj})/(n + 4)}$$

Even though the adjustment proposed by Agresti-Coull is relatively minor, it provides a significant improvement in coverage probability of the confidence interval relative to the Wald method. As shown in Figure 3.6, there are very few "unlucky" p, in the sense that the coverage probability almost uniformly exceeds the nominal 95% level for nearly all p even for a relatively small sample size. Later in the chapter, we will show that the Agresti–Coull confidence interval can also be viewed as a Bayesian confidence interval.

FIGURE 3.6
Coverage probabilities of Agresti–Coull confidence intervals for $n = 20$.

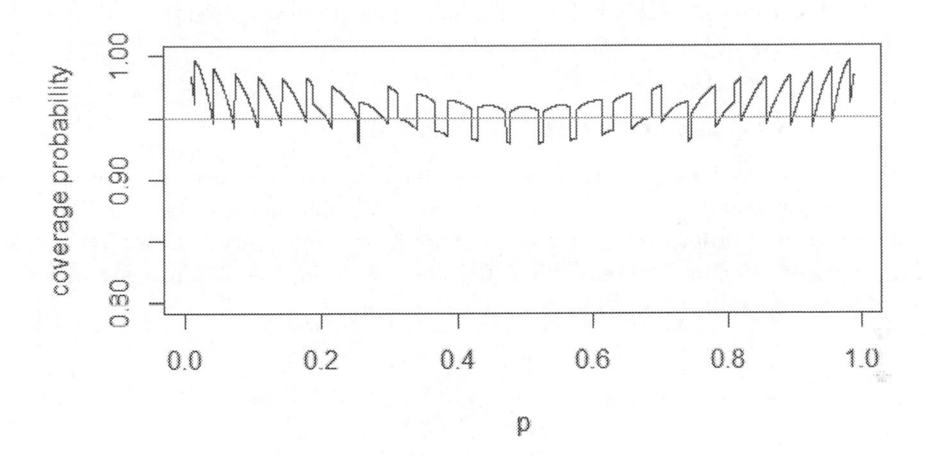

Illustration: Asymptotic interval due to Agresti and Coull (1998)

In real-world applications, this method is increasingly popular in computing confidence intervals of a binomial proportion. To illustrate this method, again consider the SUI example introduced in Section 3.3. Using SAS PROC FREQ, the Agresti and Coull (1988) method can be employed using the following SAS code:

```
ods listing;
data sui;
input trt response freq;
cards;
1 1 28
1 2 3
;
run;
proc freq data=sui ;
tables response/nocum norow binomial (cl=ac) alpha=0.05;
weight freq;
run;
```

In the above code, the SAS keyword *binomial* is for Agresti and Coull method with the option *cl=ac*. A partial output includes the following:

```
            Confidence Limits for the Binomial Proportion
                        Proportion = 0.9032

              Type              95% Confidence Limits
```

```
Agresti-Coull    0.7431         0.9744
```

In the output, the Agresti–Coull method shows 95% confidence interval as $(0.7431, 0.9744)$. Again, for testing the hypothesis, since the lower limit 0.7431 is less than 0.75, the null hypothesis cannot be rejected, and the device is declared to be inferior to the established.

3.3.6 Second-order corrected interval

This is a one-sided asymptotic confidence interval method proposed by Cai [17]. In this method, the one-sided asymptotic CI obtained by removing the first and second-order systematic bias terms from the coverage based on its Edgeworth expansion. The resulting $100(1-\alpha)\%$ one-sided confidence intervals are obtained as follows:

$$[0, \tilde{p} + z_\alpha(V(\widehat{p}) + (\gamma_1 V(\widehat{p}) + \gamma_2)n^{-1})^{1/2}n^{-1/2}]$$

$$[\tilde{p} - z_\alpha(V(\widehat{p}) + (\gamma_1 V(\widehat{p}) + \gamma_2)n^{-1})^{1/2}n^{-1/2}, 1]$$

where $\eta = (z_\alpha^2/3 + 1/6)$, $\gamma_1 = -(13z_\alpha^2/18 + 17/18)$, $\gamma_2 = (z_\alpha^2/18 + 7/36)$, and $\tilde{p} = (x + \eta)/(n + 2\eta)$.

FIGURE 3.7
Coverage probabilities of second-order corrected confidence intervals for $n = 20$.

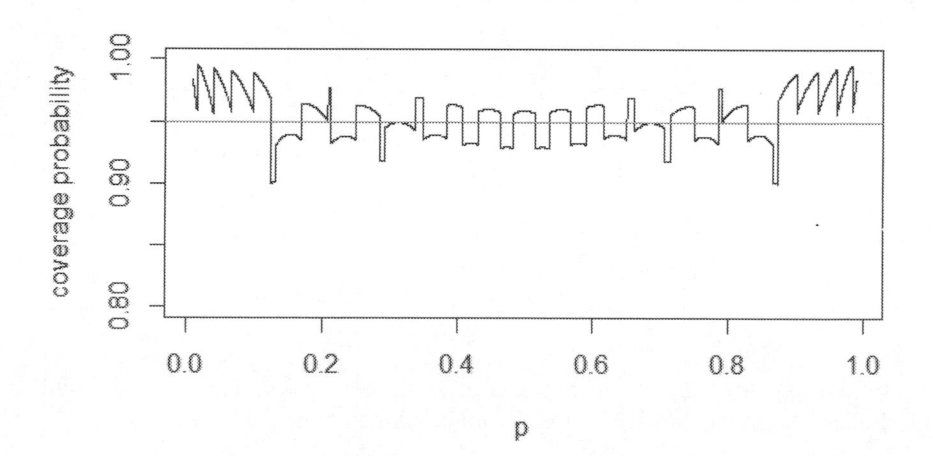

The distribution of the coverage probabilities of the second-order corrected method is shown in Figure 3.7. In terms of coverage probabilities, most asymptotic methods are vulnerable near the boundaries of the interval [0,1]; however, it appears that the second-order corrected method due to Cai [17] achieves the nominal coverage probabilities near the two extremes, namely, in the region (0,0.1) and (0.9,1). For all other p's, the coverage probabilities are comparable to the other methods, such as the Wilson score method or the Agresti and Coull method.

Illustration: Second-order corrected interval

Since this method is specifically designed for computing one-sided confidence intervals, it is beneficial for testing non-inferiority or superiority. As of SAS v9.4, this method is not available in SAS's PROC FREQ. However, it can be easily implemented with simple programming in any statistical package. The following code written in R can compute a two-sided confidence interval (adjusted for the significance level). To illustrate this method, again consider the SUI example introduced earlier in Section 3.3:

```
second.order<-function(x,n,conflev)
{
  p<-x/n

  b<--1
  k<- abs(qnorm((1-conflev)/2))

  k2<-k^2
  eta<-(1/3)*k2+(1/6)
  #print(eta)
  gamma1<-b*((13/18)*k2+(17/18))
  gamma2<-(1/18)*k2+(7/36)
  #print (gamma2)
  miu_<-(x+eta)/(n-(2*eta*b))
  Vmiu<-p*(1-p)
  CI2u<-miu_+k*((Vmiu+(gamma1*Vmiu+gamma2)/n)^0.5)*(n^(-0.5))
  CI2l<-miu_-k*((Vmiu+(gamma1*Vmiu+gamma2)/n)^0.5)*(n^(-0.5))
  c(CI2l,CI2u)
}

second.order(28,31,0.95)
```

When the above code is submitted in R, it shows the following output:

```
> second.order(28,31,0.95)
[1] 0.7631407 0.9744464
```

The R output for this method shows 95% confidence interval as $(0.7631, 0.9744)$. Since the lower limit 0.7631 is greater than 0.75, the null hypothesis is rejected, and the device is declared to be non-inferior to the established.

3.4 Bayesian intervals

As discussed in chapter 1, Bayesian intervals (known as credible intervals) are based on the Bayes theorem. In general, credible intervals can be used to carry out inference of an unknown quantity of interest, say $\boldsymbol{\theta}$, which can be a model parameter, missing data, or prediction. In the current context, consider $\boldsymbol{\theta} \equiv p$. For an observed data $\boldsymbol{y}$, let $L(\boldsymbol{y}|\boldsymbol{\theta})$ is the likelihood of the data given the unknown parameter $\boldsymbol{\theta}$. Let $\pi(\boldsymbol{\theta})$ the prior distribution that describes the uncertainties of the unknown quantities $\boldsymbol{\theta}$ independent of the current data. Hence, the posterior distribution of $\boldsymbol{\theta}$ given by

$$P(\boldsymbol{\theta}|\boldsymbol{y}) \propto L(\boldsymbol{y}|\boldsymbol{\theta})\pi(\boldsymbol{\theta}) \tag{3.9}$$

For an observed data, i.e., the total number of successes, $x \equiv \boldsymbol{y}$, $x \sim Binomial(n, \theta)$, and the likelihood $L(\boldsymbol{y}|\boldsymbol{\theta})$ is

$$L(\boldsymbol{y}|\boldsymbol{\theta}) = \binom{n}{x}\theta^x(1-\theta)^{n-x} \equiv f(x,\theta)$$

In the Bayesian framework, the identification of the posterior distribution can be challenging, and in this context, the choice of the prior distribution of $\boldsymbol{\theta}$ plays an important role. A class of prior distribution, known as the conjugate prior, is frequently used in Bayesian analysis. As discussed in chapter 1, a conjugate prior can be viewed as a special prior distribution that matches with the likelihood, in the sense that the posterior distribution belongs to the same parametric family as the prior. A conjugate prior is a convenient mathematical tool in simplifying the computation of the posterior distribution. In the above likelihood equation, the parametric form, also known as the kernel, is $\theta^x(1-\theta)^{n-x}$. The distribution that matches this kernel is the beta distribution. Consider $\pi(\boldsymbol{\theta}) \propto B(\alpha, \beta)$, where $B(\alpha, \beta)$ is the beta distribution with parameters α and β. Hence, from Equation 3.9

$$\begin{aligned} P(\boldsymbol{\theta}|\boldsymbol{y}) &\propto L(\boldsymbol{y}|\boldsymbol{\theta})\pi(\boldsymbol{\theta}) \\ &\propto \frac{\Gamma(\alpha+\beta)}{\Gamma(\alpha)\Gamma(\beta)}\binom{n}{x}\theta^{x+\alpha-1}(1-\theta)^{n-x+\beta-1} \\ &\propto \theta^{x+\alpha-1}(1-\theta)^{n-x+\beta-1} \\ &\equiv B(x+\alpha, n-x+\beta) \end{aligned} \tag{3.10}$$

Therefore, the posterior distribution of p with respect to the Beta prior $B(\alpha, \beta)$ is $B(x+\alpha, n-x+\beta)$. This observation can be the basis for constructing Bayesian credibility intervals of p, which is discussed in the following sections.

3.4.1 Jeffreys interval

A particular and interesting choice of Beta prior is B(1/2,1/2), i.e., the conjugate Beta prior with $\alpha = 1/2, \beta = 1/2$, which has a pdf $f(\theta) \propto$

$\theta^{-1/2}(1-\theta)^{-1/2}$, $0 < \theta < 1$. This is a special case of the class of non-informative priors known as the Jeffreys prior. It follows from (3.10), the two-sided equal-tailed credible interval based on the Jeffreys prior with a posterior probability of $(1-\alpha)$ is given by:

$$\underline{\Delta}(x) = B(\alpha/2; x+1/2, n-x+1/2)$$

$$\overline{\Delta}(x) = B(1-\alpha/2; x+1/2, n-x+1/2)$$

where $B(\alpha; a, b)$ denotes the α-th quantile of B(a,b) distribution.

FIGURE 3.8
Coverage probabilities of Jeffreys confidence intervals for $n = 20$.

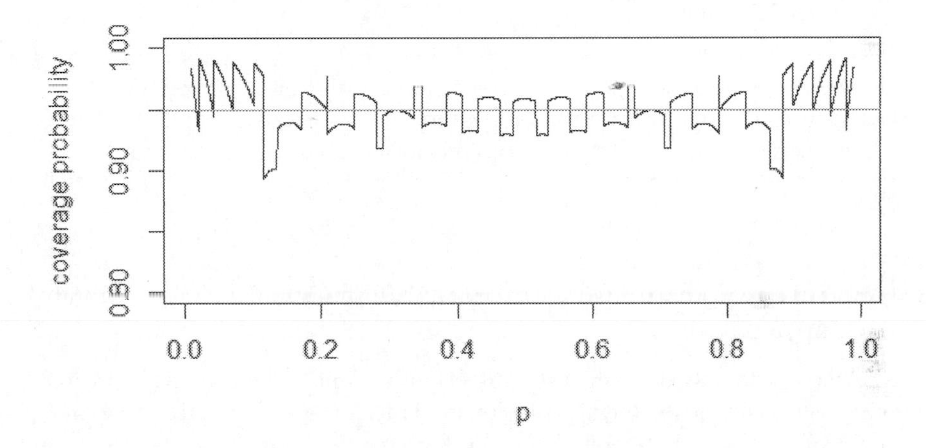

The distribution of the coverage probabilities of the Jeffreys method with $n = 20$ is shown in Figure 3.8, which is very similar to that of the second-order corrected method due to Cai [17].

Illustration: Jeffreys interval

This particular Bayesian method has been implemented in most statistical software packages and is also recommended by Brown, Cai and DasGupta [15]. To illustrate this method, consider the SUI example introduced in Section 3.3. Using SAS PROC FREQ, the Jeffreys interval can be obtained using the following SAS code:

```
ods listing;
data sui;
input trt response freq;
```

```
cards;
1 1 28
1 2 3
;
run;
proc freq data=sui ;
tables response/nocum norow binomial (cl=jeffreys) alpha=0.05;
weight freq;
run;
```

In the SAS code, the keyword *binomial (cl=jeffreys)* invokes a one-way binomial confidence interval method using Jeffreys prior. SAS's partial output includes the following:

```
          Confidence Limits for the Binomial Proportion
                      Proportion = 0.9032

               Type          95% Confidence Limits

               Jeffreys      0.7637          0.9720
```

The above output for the Jeffreys method shows the 95% confidence interval as $(0.7637, 972)$. Since the lower limit, 0.7637, is greater than 0.75, the null hypothesis is rejected, and the non-inferiority of the device is established.

3.4.2 Intervals using non–informative priors: General MCMC approach

The SAS code discussed in the last subsection computes the Bayesian credible interval using the closed-form posterior distribution given in 3.10. However, in Bayesian analysis, it is not always possible to find a closed-form solution of the posterior distribution. In such situations, a general approach to identifying posterior distribution is based on the MCMC, i.e., the Markov Chain Monte Carlo method. This is a simulation-based approach that allows random samples to be drawn from the target distribution, which can be used to approximate the posterior distribution. The method has been implemented in SAS's PROC MCMC, which provides a more versatile method to Bayesian inference. The following SAS code offers an alternative approach to obtain credible intervals for the Jeffreys prior discussed in the last subsection.

```
ods listing;
data sui;
input trt response n;
cards;
1 28 31
;
run;

proc mcmc data=sui seed=12345 nmc=1000000 stats=int;
parms p 0.5;
```

```
prior p~beta(0.5,0.5);
llike=logpdf('binomial',response,p,n);
model general(llike);
run;
```

In the SAS code, the data step has been slightly altered, and has been written as the binomial data (specifying 28 success out of 31 trials). This change is needed as in the PROC MCMC log-likelihood has been specified as "binomial". Also, note that in the *PROC MCMC* keyword *seed=12345 nmc=1000000 stats=int* is for a fixed seed 12345, nmc=1000000 is for the MCMC samples, and the stats=int invokes equal-tailed and HPD (highest posterior density) credible intervals (see chapter 1 for a discussion on the HPD interval). The keyword *parms p 0.5* initializes the parameter p, and the syntax *prior p* $\sim$ *beta(0.5,0.5)* specifies beta(0.5, 0.5) prior. The data has been passed as binomial log-likelihood and specified in the syntax *llike=logpdf("binomial",response,p,n)*. Finally, the syntax *model general(llike)* is for the general log-likelihood, a standard syntax in PROC MCMC. The diagnostic plots shown in Figure 3.9 from PROC MCMC indicate no issue with the Markov chains.

FIGURE 3.9
Bayesian diagnostics for a proportion from PROC MCMC.

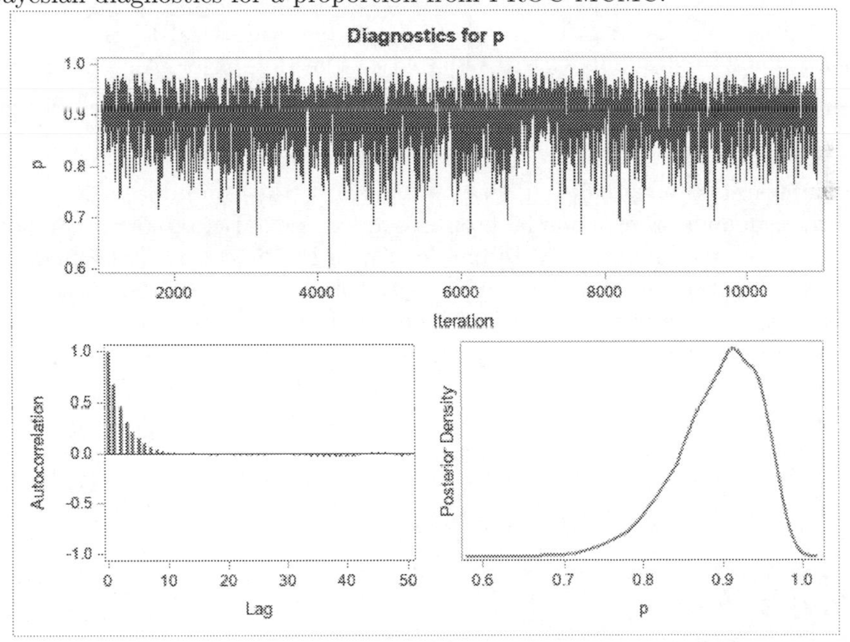

The top panel of Figure 3.9 is known as the trace plot, which shows the convergence and mixing of the chain from the MCMC sampler of SAS implementation that explores the parameter space. Details of the MCMC method are not given here; however, any introductory book on Bayesian analysis would include extensive discussion of the methodology. The trace plot shows that the chain is stabilized and randomly distributed about a center of 0.9, which is close to the posterior mean. The MCMC samples are generally correlated, and a significant serial correlation is a cause of concern. There are a number of ways to reduce correlation, such as a method called "thinning," in which some samples are systematically discarded. The bottom panel of Figure 3.9 contains two plots – the left bottom plot showing the autocorrelation of the MCMC samples. The correlation on the y-axis is plotted against the lag shown on the x-axis. From this plot, it is clear that the autocorrelation decays rapidly and disappears after a lag of 10 or so, which suggests that in this context, the level of autocorrelation is not a cause of concern. The bottom right panel shows the smooth posterior density plot.

A partial output of SAS PROC MCMC includes the following:

```
                        Posterior Intervals

Parameter    Alpha     Equal-Tail Interval         HPD Interval

p            0.050      0.7637       0.9719     0.7835       0.9817
```

Note that in the 95% HPD interval and the equal-tail interval for p are given by $(0.7835, 0.9817)$ and $(0.7637, 0.9719)$, respectively. The credible interval based on the MCMC distribution is almost identical to the Jeffreys interval computed in PROC FREQ based on the Beta prior. Also, note that the HPD interval computed from the MCMC distribution is narrower than the equal-tail credible interval, which is expected as discussed in chapter 1.

In the example Beta (0.5, 0.5) has been used as the non-informative prior. In Bayesian analysis of Binomial proportion, the use of Beta (0.5, 0.5) or Beta (1, 1) priors are common and recommended in the literature. Note that the Beta (1,1) is the same as the Uniform (0,1) distribution. In the above SAS code, one can alternatively specify Uniform (0,1) prior following the simple adjustment to the prior statement:

```
ods listing;
data sui;
input trt response n;
cards;
1 28 31
;
run;
ods graphics on;
proc mcmc data=sui seed=12345 nmc=1000000 stats=int;
parms p 0.5;
prior p~uniform(0,1);
llike=logpdf('binomial',response,p,n);
model general(llike);
run;
```

As before, the output includes several diagnostic plots. The Posterior intervals are as follows:

```
Parameter    Alpha      Equal-Tail Interval          HPD Interval

p            0.050      0.7503      0.9650      0.7688      0.9750
```

Note that these intervals are roughly comparable to the intervals based on Jeffreys prior discussed earlier, but not exactly the same, which is due to the effect of prior on Bayesian computation.

3.4.3 Intervals using informative priors: Power prior

The previous section briefly describes the Bayesian methods using non-informative priors. This section introduces a popular method to Bayesian inference based on prior distributions known as power priors (see Ibrahim and Chen (2000)), which can be viewed as an informative prior. In clinical trials and many other applications, often historical data are available, and it may be useful to incorporate the information via a prior distribution in Bayesian data analysis of data. The power priors are a family of priors that easily incorporate historical data relevant to the current analysis. Let $D \equiv (n, \boldsymbol{y}, \boldsymbol{x})$ and $D_0 = (n_0, \boldsymbol{y}_0, \boldsymbol{x}_0)$ are the current data and the historical data, respectively, then the power prior is defined as

$$\pi(\theta|D_0, a_0) \propto L(\theta|D_0)^{a_0}\pi_0(\theta) \tag{3.11}$$

where $0 \leq a_0 \leq 1$ is called discount parameter, $L(\theta|D_0)$ is the likelihood of θ using the historical data, and $\pi_0(\theta)$ is the prior distribution of θ before any data is observed. Therefore, the term $L(\theta|D_0)^{a_0}$ is the weighted likelihood of θ given historical data. Using Equation 3.9, and given the observed data $D \equiv (n, \boldsymbol{y}, \boldsymbol{x})$, the posterior $P(\theta|D, D_0, a_0)$ can be written as

$$\begin{aligned} P(\theta|D, D_0, a_0) &\propto L(\theta|D)\pi(\theta|D_0, a_0) \\ &\propto L(\theta|D)L(\theta|D_0)^{a_0}\pi_0(\theta) \\ &\propto \left(\prod_{i=1}^{n} f(y_i|\theta, x_i\right)\left(\prod_{i=1}^{n_0} f(y_{0i}|\theta, x_{0i}\right)^{a_0}\pi_0(\theta) \end{aligned} \tag{3.12}$$

where $f(.,.)$ is the likelihood function for a single observation in either historical data or the current data.

Illustration: Power prior using PROC MCMC

To illustrate this method, consider the SUI data introduced earlier in Section 3.3. Assume that there is some historical data of the same device treating SUI where 85 of 97 patients reported success. If the sponsor is interested in incorporating historical information into the current data, this can be easily accomplished using power prior. However, let's assume that the sponsor does not want to put too much weight on the historical data, and it is decided apriori that only 20% weight will be given on the historical data. To implement the Bayesian analysis using power prior, let us first change the SUI data slightly using the following code where *group=1* is introduced for the current data and *group=2* for the historical data. Finally, the two data are combined using the following data steps:

```
/*------current data, group=1--------------*/
data sui;
input group response n;
cards;
1 28 31
;
/*------historical data, group=2--------------*/
data hist;
input group response n;
datalines;
2 85 97
;

data combo;
set sui hist; run;
```

TABLE 3.1
Combo data structure in SAS.

group	response	n
1	28	31
2	85	97

The combined data in SAS (in the above code, it is called *Combo*) is given in Table 3.1, which is utilized in the following *PROC MCMC* code:

```
ods listing;
ods graphics;
proc mcmc data=combo seed=12345 nmc=100000 stats=int outpost=pout;
parm p 0.5;
/*discount parameter a0=0.2 is specified below*/
begincnst;
a0 = 0.2;
endcnst;
prior p ~ uniform(0, 1);
llike = logpdf(``binomial", response, p, n);
if (group =2) then
llike = a0 * llike;
model general(llike);
run;
```

PROC MCMC shows the diagnostic plots given in Figure 3.10. The upper panel displays the traceplot, which confirms the convergence of the MCMC chain in exploring the posterior distribution. The bottom left panel's autocorrelation plot shows that the serial correlation of the MCMC samples decays to 0 after lag=15 or so, and the bottom right panel displays the smooth posterior density. The diagnostic plots suggest no significant issues with MCMC samples.

FIGURE 3.10
Bayesian diagnostics for a proportion using power prior from PROC MCMC.

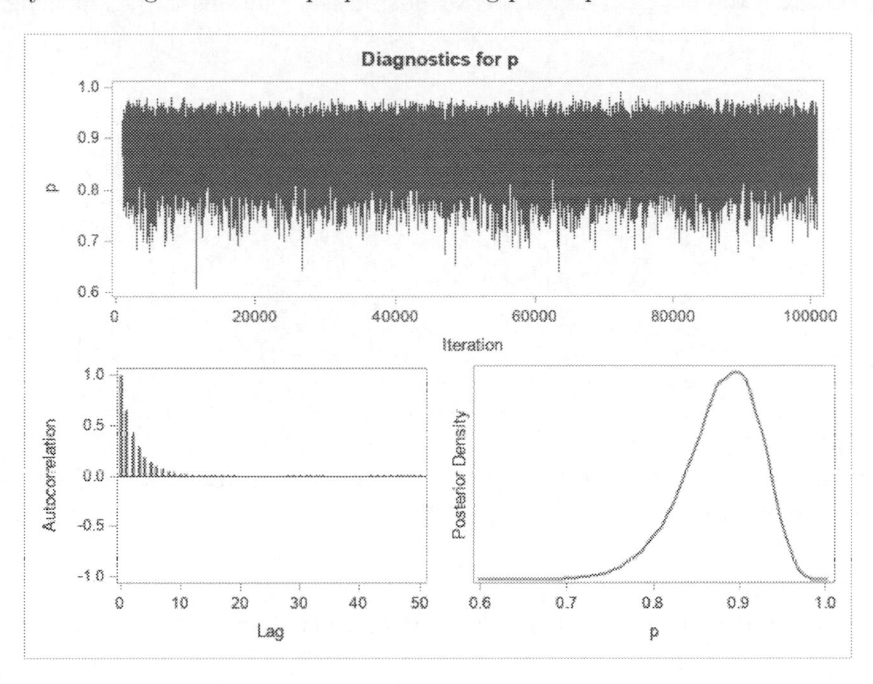

The partial output from the SAS code gives the following credible intervals.

```
                    Posterior Intervals

Parameter    Alpha    Equal-Tail Interval      HPD Interval

p            0.050    0.7763     0.9513     0.7892     0.9587
```

These intervals in this context are essentially the same as the ones obtained under non-informative priors.

3.5 Exact methods

3.5.1 Clopper and Pearson confidence interval

The Clopper–Pearson intervals are computed from the cumulative probability of Binomial distribution. It is an "exact" method in the sense that the interval is based on the exact distribution and not based on an approximation to the Binomial distribution. The approach involves solutions to two equations that require the upper and lower tail probabilities of Binomial distribution is $\alpha/2$, which ensures at least nominal coverage for both two-sided and one-sided CI's. To obtain the limits $\underline{\Delta}(x)$ and $\overline{\Delta}(x)$, we look for the solutions to the following:

$$P(X \geq x | \underline{\Delta}(x), n) = \alpha/2$$

$$P(X \leq x | \overline{\Delta}(x), n) = \alpha/2$$

The limits can also be obtained from the relationship between Binomial, Beta, and F-distribution (Collett [29], Leemis and Trivedi [54]), and is given by:

$$\underline{\Delta}(x) = \left[1 + \frac{n - x + 1}{x F_{2x, 2(n-x+1), \alpha/2}}\right]^{-1}$$

$$\overline{\Delta}(x) = \left[1 + \frac{n - x}{(x+1) F_{2(x+1), 2(n-x), 1-\alpha/2}}\right]^{-1}$$

where $F_{a,b,\alpha}$ denotes the upper 100α percentile point of the F distribution with a and b degrees of freedom. The lower and upper limits of the interval are set at 0 and 1 for $x = 0$ and $x = 1$, respectively.

FIGURE 3.11
Coverage probabilities of exact Clopper–Pearson confidence intervals for $n = 20$.

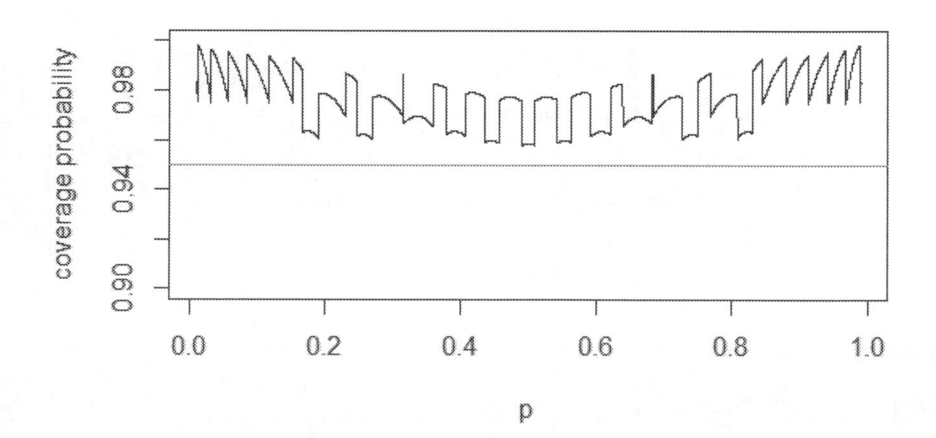

The nominal coverage probabilities of the exact Clopper–Pearson method are shown in Figure 3.11. Note that all coverage probabilities are well above the

95% level guaranteeing the nominal coverage. Therefore, when confidence intervals for small samples are required in real-world applications, the Clopper-Person method is always recommended. One drawback of using such a method is that the confidence intervals are conservative, i.e., the confidence intervals are wider than the same using the asymptotic methods.

Illustration: Exact Clopper–Pearson confidence interval

Consider the SUI example introduced earlier in Section 3.3. Using SAS PROC FREQ, the exact Clopper–Pearson confidence interval can be computed using the following SAS code:

```
ods listing;
data sui;
input trt response freq;
cards;
1 1 28
1 2 3
;
run;
proc freq data=sui ;
tables response/nocum norow binomial (cl=exact) alpha=0.05;
weight freq;
run;
```

In the code, the SAS keyword *binomial (cl=exact)* is for the Clopper–Person method using exact distribution where the keyword *(cl=exact)* invokes the Clopper–Pearson method. The partial output includes the following:

```
Confidence Limits for the Binomial Proportion
              Proportion = 0.9032

Type                         95% Confidence Limits

Clopper-Pearson (Exact)      0.7425          0.9796
```

The SAS output for Clopper–Pearson shows 95% confidence interval as $(0.7425, 0.9796)$. Since the lower limit 0.7425 is less than 0.75, the null hypothesis cannot be rejected, and the inferiority of the device is established.

3.5.2 Mid-p corrected Clopper–Pearson interval

As stated in the last section, the confidence intervals using the Clopper–Pearson method are conservative. The Mid-p corrected Clopper–Pearson is a method of computing a less conservative version of Clopper–Pearson confidence interval (see Agresti and Gottard [4]). This interval cannot be obtained in closed form but can be computed by numerically solving the equations:

$$P(X \geq x|\underline{\Delta}(x), n) - \frac{1}{2}P(X = x|\underline{\Delta}(x), n) = \alpha/2$$

$$P(X \leq x|\overline{\Delta}(x), n) - \frac{1}{2}P(X = x|\overline{\Delta}(x), n) = \alpha/2$$

FIGURE 3.12
Coverage probabilities of mid-p corrected Clopper–Pearson confidence intervals for $n = 20$.

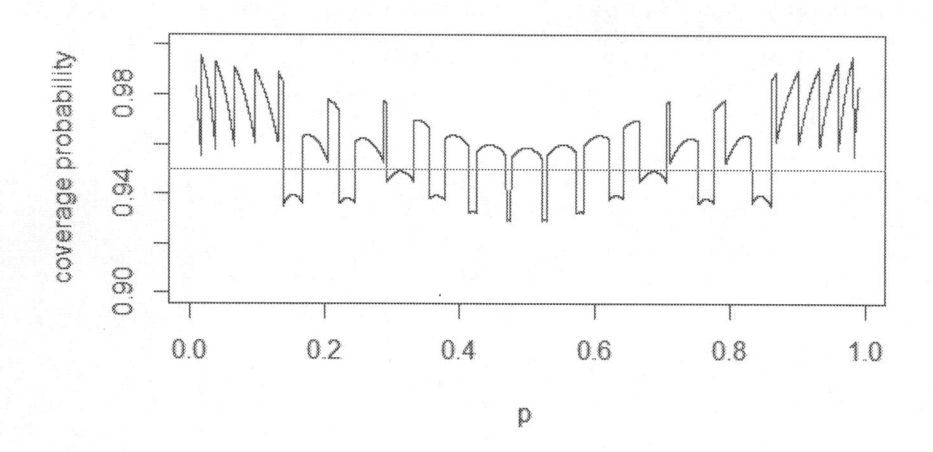

where $\underline{\Delta}(x) = 0$ if $x = 0$, and $\overline{\Delta}(x) = 1$ if $x = n$. Figure 3.12 shows the distribution of the coverage probabilities using mid-p corrected Clopper–Pearson method. As shown in the figure, even though it reduces conservativeness of the Clopper–Pearson method (some of the confidence intervals are less wider than the same using the Clopper–Pearson method), in some situations, it may not achieve the nominal coverage level.

Illustration: Mid-p corrected Clopper–Pearson confidence interval

Consider the SUI example introduced earlier in Section 3.3. Using SAS PROC FREQ, the exact confidence interval using Mid-p corrected Clopper–Pearson can be computed using the following SAS code:

```
ods listing;
data sui;
input trt response freq;
cards;
1 1 28
1 2 3
;
run;
proc freq data=sui ;
tables response/nocum norow binomial (cl=midp) alpha=0.05;
weight freq;
run;
```

The SAS keyword *binomial (cl=midp)* is for Mid-P corrected Clopper–Person method using exact distribution. The partial SAS output gives the following result:

```
Confidence Limits for the Binomial Proportion
             Proportion = 0.9032

        Type      95% Confidence Limits

        Mid-p     0.7588           0.9748
```

The SAS output for the Mid-p corrected Clopper–Pearson reports 95% confidence interval as $(0.7588, 0.9748)$. Since the lower limit 0.7588 is greater than 0.75, the null hypothesis is rejected, and the non-inferiority of the device is established.

3.5.3 Confidence interval due to Casella

This confidence interval method is available only in StatXact software. A confidence set of p, say C, is a collection of $(n+1)$ intervals,$(\underline{\Delta}(x), \overline{\Delta}(x))$, $x = 0, 1, ..., n$. Thus, the coverage of the confidence procedure is given by

$$\inf_p P(p \in C|p, n) = \inf_p \left[\sum_{x=0}^{n} I_{(\underline{\Delta}(x), \overline{\Delta}(x))}(p) \binom{n}{x} p^x (1-p)^{(n-x)} \right] \geq 1 - \alpha$$

The expression inside the parentheses represents the probability that the random interval $(\underline{\Delta}(x), \overline{\Delta}(x))$ contains the true parameter value p. Also, note that if

$$\inf_p P(p \in C|p, n) \geq 1 - \alpha$$

for a specified minimum coverage $1-\alpha$, it is possible to construct a refined procedure $C^* = \{(\underline{\Delta}^*(x), \overline{\Delta}^*(x)); x = 0, 1, ..., n$ directly by increasing $\underline{\Delta}(x)$ to $\underline{\Delta}^*(x)$ (and decreasing $\overline{\Delta}(x)$ by the same amount) starting from $x = n$ until we reach $P(\underline{\Delta}^*(x) \in C|\underline{\Delta}^*(x), n) = 1 - \alpha$ for all x. Casella [20] gives an algorithm to produce a collection of refined intervals with uniformly shorter lengths when applied to any $1-\alpha$ binomial confidence procedure. The resulting confidence intervals have some optimal properties (Casella [20]). As noted by Newcombe [63] (cf. Table 6), shortened intervals like Sterne's [87] have erratic location properties. Consequently, its impact on the coverage of one-sided CI is worth exploring. Our simulation study (Section 3) clearly shows that the one-sided CI's based on two-sided BL (Blaker) and BS (Blyth–Still–Casella) intervals suffer from serious undercoverage even for very large sample sizes. Figure 3.13 shows the distribution of the coverage probabilities of Casella (1986) method. As shown in the figure, the confidence intervals using the exact Casella method consistently achieve a nominal coverage level. Among all other exact methods, this method is known to be less conservative, hence recommended in most real-world applications.

FIGURE 3.13
Coverage probabilities of Casella confidence intervals for $n = 20$.

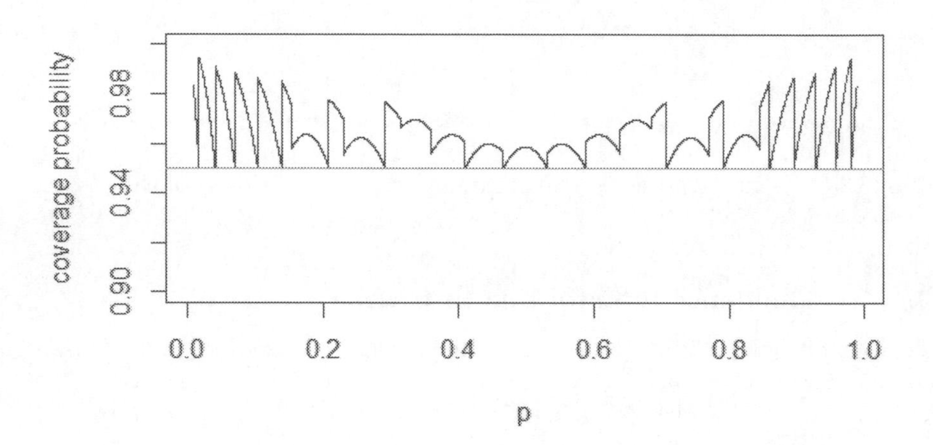

Illustration: Exact confidence interval due to Casella (1986)

Consider the SUI example introduced earlier in Section 3.3. As stated before, this method is available in StatXact software. Using the SAS version of the StatXact PROCs software, the exact confidence interval using the Casella method (in StatXact it is referred to as Blyth–Still–Casella method) can be computed using the following code:

```
data sui;
input trt response freq;
cards;
1 1 28
1 2 3
;
run;
proc binomial data=sui ;
BI/BS ;
OU response;
weight freq;
run;
```

In the code, *proc binomial*, part of StatXact PROCs, invokes a SAS procedure specially developed for SAS software. The keyword *BI/BS* is for Casella's (1986) method to compute a confidence interval. By default PROC BINOMIAL computes a 95% confidence interval; *OU response* is for the outcome (*OU* is the keyword) variable *response* where the variable *response* has to be passed in as a binary variable with 1 as the success. The syntax *weight freq* is for the usual weight variable for the response variable. StatXact PROCs partial output includes the following:

```
ESTIMATION OF BINOMIAL PARAMETER (PI)

Data file name : < SUI >
Outcome Variable Name : response
Weight Variable Name : freq


    Number of Trials          = 31
    Number of Successes       = 28

    Maximum Likelihood Estimate of PI  =      0.9032

     95.00% Confidence Interval for PI:
            (Clopper-Pearson)      = (      0.7425 ,       0.9796)
Mid-P corrected (Clopper-Pearson) = (      0.7588 ,       0.9748)
            (Blyth/Still/Casella) = (      0.7545 ,       0.9731)
```

The output corresponding to *(Blyth/Still/Casella)* is for Casella (19986) method, and it shows the 95% confidence interval as $(0.7545, 0.9731)$. Again, for testing the hypothesis, since the lower limit 0.7545 is greater than 0.75, the null hypothesis is rejected, and the non-inferiority of the device is established.

3.5.4 Confidence interval due to Blaker

The Clopper–Person method introduced in the earlier section is conservative and tends to give wider confidence intervals. The exact confidence interval due to Blaker (2000), on the other hand, is less conservative and hence increasingly popular in real-world applications. Blaker uses Spjøtvoll's (Spjøtvoll [9], Blaker and Spjøtvoll [12]) notion of a preference function (PF) to improve upon the interval proposed by Birnbaum [10]. Birnbaum bases his interval on the P-value of the equal-tailed test of $H_0 : p = p_0$ as follows:

Given x, $B_\alpha(x) = \{p : \beta(p; x) > \alpha\}$ is a CI with minimum coverage $1 - \alpha$, where $\beta(p; x) = min[2min\{P(X \geq x|p, n), P(X \leq x|p, n)\}, 1]$.

Note $\beta(p; x)$ represents the p-value of the equal-tailed test. Spjøtvoll defines PF $\beta(p; x)$ as a real-valued function on the parameter space for each observed x. Given a PF $\beta(p; x)$ and an observed x, the parameter value p_1 is preferable to p_2 if $\beta(p_1; x) > \beta(p_2; x)$. Note that $\beta(p; x)$ given by Birnbaum is a PF and has the additional property that $\{p : \beta(p; x) > \alpha\}$ is a $1 - \alpha$ CI for p. Blaker considers the PF $\lambda(p; x) = P\{\gamma(p; X) \leq \gamma(p; x)\}$, where $\gamma(p; x) = min\{P(X \geq x), P(X \leq x)\}$, and proves that the set $C_\alpha(x) = \{p : \lambda(p; x) > \alpha\}$ is a CI with minimum coverage $1 - \alpha$. Since $\beta(p; x) \geq \lambda(p; x)$(cf. Blaker [11]) for all x, the CI's based on $\lambda(p; x)$ are shorter than the ones based on $\beta(p; x)$ and hence is an improvement over Birnbaum's interval.
Figure 3.14 shows the distribution of the coverage probabilities of the Blaker method. As shown in the figure, the confidence intervals using the Blaker approach consistently achieve a nominal coverage level. Among all other exact methods, this is known to be less conservative, hence recommended in most real-world applications.

FIGURE 3.14
Coverage probabilities of Blaker confidence intervals for $n = 20$.

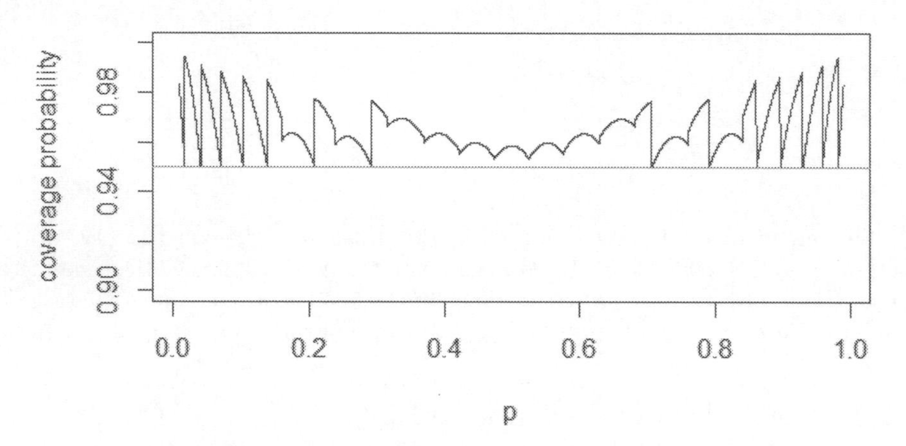

Illustration: Exact interval due to Blaker (2000)

Consider the SUI example introduced in Section 3.3. Using SAS PROC FREQ in SAS 9.4, the exact confidence interval due to Blaker can be computed using the following SAS code:

```
ods listing;
data sui;
input trt response freq;
cards;
1 1 28
1 2 3
;
run;
proc freq data=sui ;
tables response/nocum norow binomial (cl=blaker) alpha=0.05;
weight freq;
run;
```

In the above, the SAS keyword *binomial (cl=blaker)* is for the Blaker method. The partial output includes the following:

```
Confidence Limits for the Binomial Proportion
            Proportion = 0.9032

       Type      95% Confidence Limits

       Blaker    0.7489          0.9731
```

The SAS output using Blaker shows 95% confidence interval as $(0.7489, 0.9731)$. Since the lower limit, 0.7489, is slightly smaller than 0.75, the null hypothesis fails to be rejected, and the device's inferiority is established.

3.6 Discussion and recommendation

This chapter includes most interval estimation methods available in the literature for Binomial proportion p and illustrates these approaches using the SUI data introduced in Section 3.3. Except for the second-order corrected interval (Cai [17]) and Casella [20], all of these methods are available in SAS's PROC FREQ. In PROC FREQ, one can run the following simple syntax to explore different methods.

```
ods listing;
data sui;
input trt response freq;
cards;
1 1 28
1 2 3
;
run;
proc freq data=sui ;
tables response/nocum norow binomial (cl=wald(correct)
score(correct) ac jeffreys exact midp blaker) alpha=0.05;
weight freq;
run;
```

and get the following partial output (sorted alphabetically by methods):

Confidence Limits for the Binomial Proportion
Proportion = 0.9032

Type	95% Confidence Limits	
Agresti Coull	0.7401	0.9744
Blaker	0.7489	0.9731
Clopper-Pearson (Exact)	0.7425	0.9796
Jeffreys	0.7637	0.9720
Mid-p	0.7588	0.9748
Wald (Corrected)	0.7830	1.0000
Wilson (Corrected)	0.7310	0.9747

As evident from the above output, when an inference is derived from these methods (especially using a small sample), a statistician may arrive at conflicting conclusions. Therefore, a careful investigation of the performances of these methods is essential. For binomial confidence interval, to evaluate the confidence interval's performance, one has to consider the nominal coverage probability and the expected length as stated in Brown et al. [15], where the nominal coverage probability and expected length are computed using the following equations:

$$\zeta(p:n) = \sum_{x=0}^{n} P(X = x|n,p) I(p \in (\underline{\Delta}, \overline{\Delta})) \tag{3.13}$$

$$\lambda(p:n) = \sum_{x=0}^{n} P(X = x|n,p)(\overline{\Delta} - \underline{\Delta}) \tag{3.14}$$

Note the nominal coverage (or simply noted as the coverage) and expected length are functions of success probability p and sample size n. For a fixed sample size, the distributions of these entities could be found assuming a uniform distribution of p.

The rest of this section presents simulation results evaluating the performance of eight most frequently used methods in practical applications, namely, Bayesian method Jeffreys (JF) using non-informative prior B(0.5, 0.5) as recommended by Brown et al. [15], Agresti and Coull method (AC), Wilson's method (SC, as known as Score method), continuity corrected Wilson (SCC) method, Casella (1986) method due to Blyth–Still (BS), Blaker method (BL), Clopper–Pearson method (CP), and Mid-p corrected Clopper–Pearson method. The performances of the Wald method or the Wald with continuity correction method are not included since their performances are way below the others on this list; per Vollset [95], *"use of the standard textbook method* $x/n \pm 1.96\sqrt{((x/n)(1-x/n)/n)}$*, or its continuity corrected version, are strongly discouraged"*. The second-order corrected (SOC) method proposed by Cai [17] is a one-sided method, can be easily adjusted for two-sided applications. The performance of the one-sided confidence interval of SOC method has been compared in Pradhan et al.[70]. Since the performance of two-sided confidence interval is expected to be similar (except the coverage properties of few exact methods) to that of one-sided confidence intervals, it is not included in the following simulations for two-sided confidence intervals.

For sample sizes $n = 25$ and 50, all simulations are carried out for 95% two-sided confidence intervals. To find coverage and expected length of a confidence interval, the parameter space [0, 1] is partitioned into $10,000$ equally spaced points; hence each partition point corresponds to a value of p. For each of these $10,000$ partition points, the coverage and expected lengths are computed using Equation 3.13 and Equation 3.14, respectively. The distribution of these $10,000$ points is presented using BliP plots (Lee and Tu [53]). The vertical bars in each plot show the deciles of the corresponding distribution, and the long vertical line in the coverage plot is for the 95% coverage level.

It is worth noting that in Figures 3.15 and 3.16, among the Bayesian and asymptotic methods, all methods except the SCC suffer from undercoverage. When the expected length (also known as precision) is considered, among these methods, AC and SCC are the best (though SCC has wider precision). Among the exact methods, CP and BL attain the nominal coverage levels; even though based on the method of construction of the BS intervals, the confidence interval due to BS is supposed to achieve nominal coverage probability. However, there were few partition points (for $n = 50$) where the nominal coverages of BS were not achieved, but this may be due to the software implementation of this method. The MCP method is less conservative than the CP method; however, this method fails to achieve nominal coverage for larger sample sizes. When the precision is considered along with the coverage, the BL method appears to be the best method, followed by BS and CP.

FIGURE 3.15
Coverage probabilities of binomial confidence intervals for $n = 25, 50$.

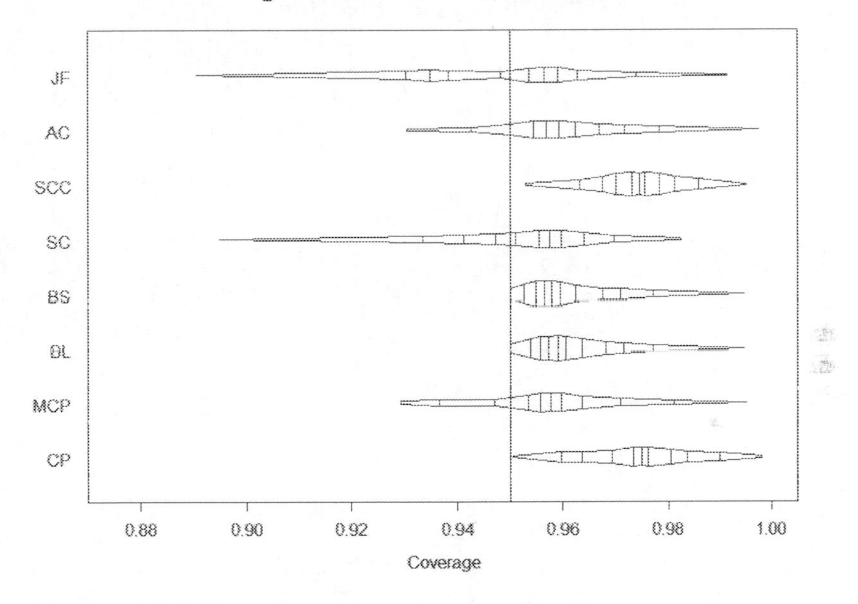

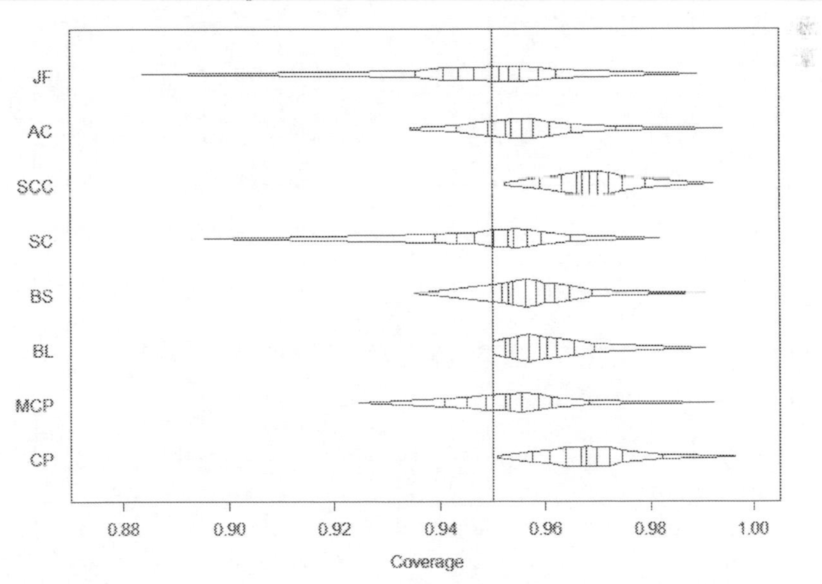

FIGURE 3.16
Expected lengths of binomial confidence intervals for $n = 25, 50$.

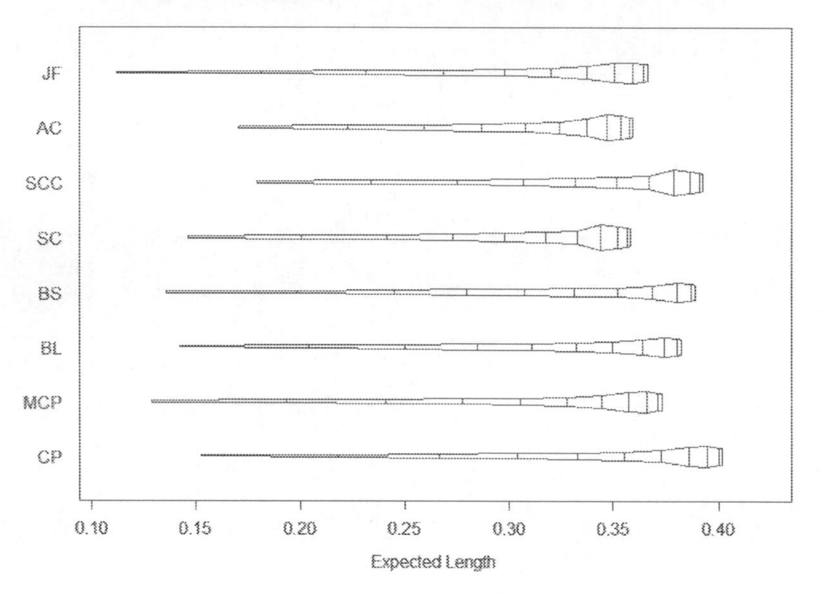

Expected lengths of 95% Nominal Intervals for n =50

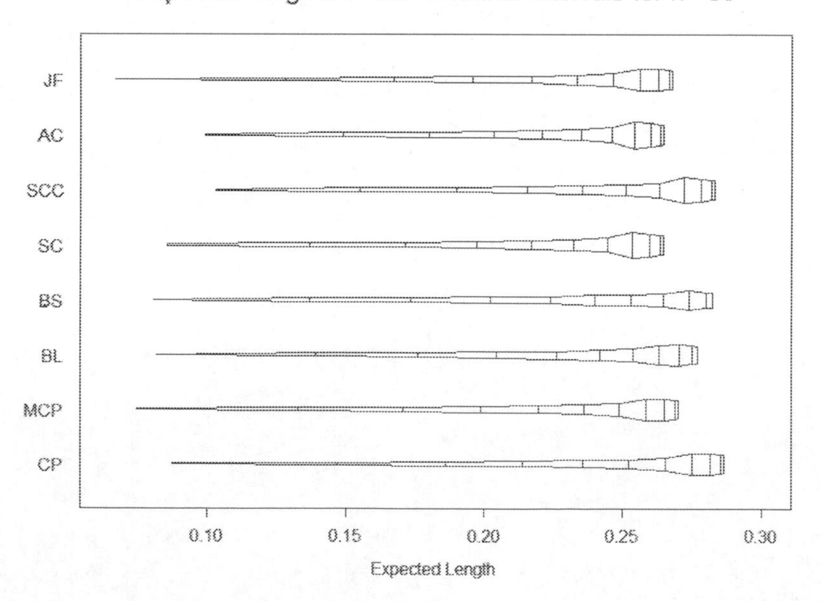

As demonstrated in Figure 3.15, exact methods are recommended in the small sample setup. Among the exact methods, BS and BL are constructed inverting two-sided tests. Hence, adjusting the nominal level and constructing a one-sided confidence interval may not have the desired coverage probability. For a one-sided interval of the form $(max(0, \underline{\Delta}(x)), 1)$ derived from two-sided interval for $n = 25$ and 50 are given in Figure 3.17.

Note that in Figure 3.17, SCC outperforms AC in terms of coverage. Among the exact methods, CP is the only method that achieves the coverage probability, hence recommended when one-sided confidence is required. In this connection, note that the performances of 1-sided intervals are not discussed in this chapter. Pradhan et al. [70] examined the small sample performances of several one-sided confidence interval methods and endorsed the CP method for all practical applications.

FIGURE 3.17
Distribution of the one-sided coverage probabilities of the binomial confidence intervals for n = 25, 50.

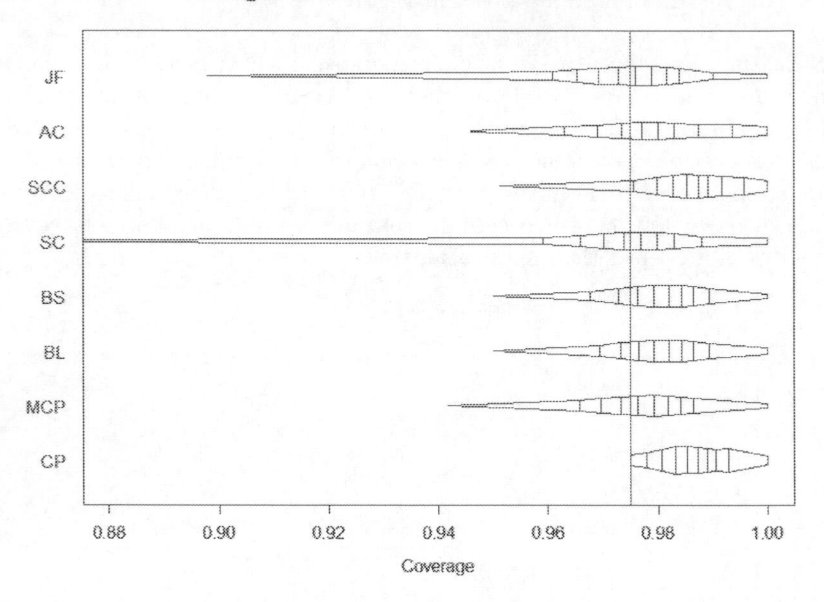

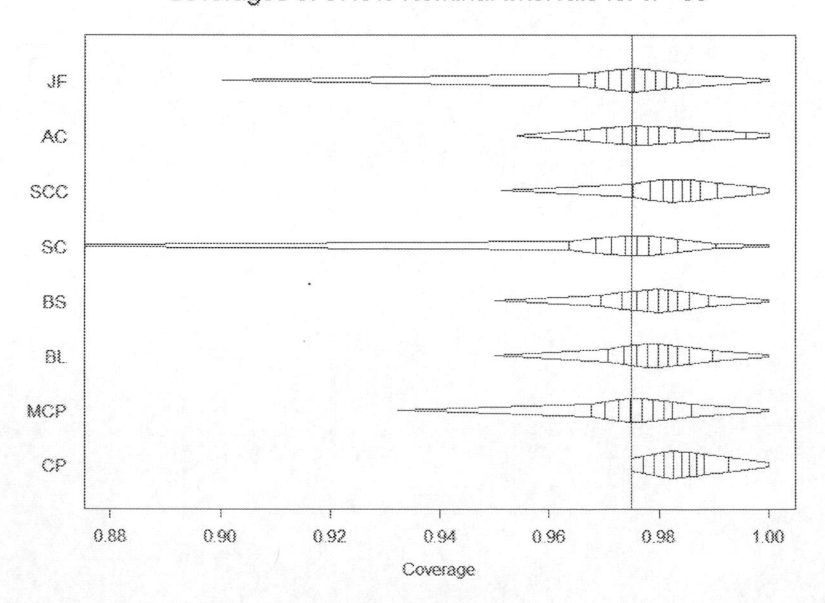

4

Two Independent Binomials: Difference of Proportions

4.1 Introduction

This chapter discusses most of the commonly used confidence intervals for the difference of two independent binomial proportions, say Δ. These intervals are used to test important hypotheses about Δ that arise in practice, especially in the context of clinical trials. These confidence intervals are either asymptotic or exact depending on their methods of construction. Real-life data from clinical trials are used to illustrate applications of these intervals in testing relevant hypotheses. The SAS codes for the computation of the intervals are provided. The chapter concludes with a brief discussion on the performances of these intervals measured in terms of their nominal coverage and expected lengths.

4.2 Difference of two proportions: $p_1 - p_2$

Let X_1 and X_2 be two independent random variables, $X_1 \sim \text{Binomial}(n_1, p_1)$ and $X_2 \sim \text{Binomial}(n_2, p_2)$. The data are typically represented in a 2×2 table as shown below:

TABLE 4.1

	Bin (n_1, p_1)	**Bin** (n_2, p_2)
Successes	x_1	x_2
Failures	$n_1 - x_1$	$n_2 - x_2$
Total	n_1	n_2

In Table 4.1, x_1 and x_2 denote the observed values of X_1 and X_2, respectively. Let $\Delta = p_1 - p_2$ and thus, $-1 \le \Delta \le 1$. A confidence interval of Δ with confidence level $(1 - \alpha)$, denoted by $(\underline{\Delta}, \overline{\Delta})$, would satisfy the following properties:

DOI: 10.1201/9781315169859-4

1. For each $\boldsymbol{x} = (x_1, x_2)$, with $0 \le x_1 \le n_1$, $0 \le x_2 \le n_2$, and for all Δ
$$P(\underline{\Delta} \le \Delta \le \overline{\Delta}) \ge 1 - \alpha. \tag{4.1}$$
2. Confidence intervals are invariant with respect to the population labels. That is, if $(\underline{\Delta}, \overline{\Delta})$ is the confidence interval of $p_1 - p_2$, then the confidence interval of $p_2 - p_1$ is
$$(-\overline{\Delta}, -\underline{\Delta}) \tag{4.2}$$
3. Confidence intervals are invariant with respect to the "success" and "failure" labels. In other words, if $(\underline{\Delta}(n_1 - x_1), \overline{\Delta}(n_2 - x_2))$ is the confidence interval of Δ based on failures $(n_1 - x_1, n_2 - x_2)$, and $(\underline{\Delta}, \overline{\Delta})$ is the confidence interval of Δ based on successes (x_1, x_2), then
$$(\underline{\Delta}(n_1 - x_1), \overline{\Delta}(n_2 - x_2)) = (-\overline{\Delta}, -\underline{\Delta}). \tag{4.3}$$
4. For a fixed sample size $\boldsymbol{n}$, all confidence intervals should be monotonic. For a fixed x_1, both limits $\underline{\Delta}(x_1, x_2)$ and $\overline{\Delta}(x_1, x_2)$ should be non-increasing in x_2, and similarly for a fixed x_2, $\underline{\Delta}(x_1, x_2)$ and $\overline{\Delta}(x_1, x_2)$ should be non-decreasing in x_1.

4.2.1 Hypotheses testing problems related to the difference of proportions

Given $\Delta_0 > 0$, a prespecified value of Δ, the common hypotheses testing problems of interest are:

1. Two-sided hypotheses testing problem:
$$H_0 : \Delta = \Delta_0 \quad \text{versus} \quad H_1 : \Delta \ne \Delta_0. \tag{4.4}$$
2. One-sided hypotheses testing problem:
$$H_0 : \Delta \le \Delta_0 \quad \text{versus} \quad H_1 : \Delta > \Delta_0 \tag{4.5}$$
or
$$H_0 : \Delta \ge \Delta_0 \quad \text{versus} \quad H_1 : \Delta < \Delta_0. \tag{4.6}$$

In clinical trials, one-sided tests are frequently used for testing the **non-inferiority** or the **superiority** of a treatment. In testing for **non-inferiority**, the objective is to test whether a drug is not inferior to a standard drug by a prespecified margin, say, Δ_0. In the following hypotheses testing problem, rejection of the null hypothesis naturally leads to the acceptance of non-inferiority of the drug,
$$H_0 : p_1 - p_2 \le -\Delta_0 \quad \text{versus} \quad H_1 : p_1 - p_2 > -\Delta_0 \tag{4.7}$$

or

$$H_0 : \Delta \leq -\Delta_0 \quad \text{versus} \quad H_1 : \Delta > -\Delta_0 \tag{4.8}$$

where p_1 and p_2 are the proportions for the test drug/population and standard drug/populations, respectively.

In equation 4.7, if $\Delta_0 < 0$, rejection of the null hypothesis leads to the acceptance of superiority (see Chow, Shao and Wang [25]) of the drug. Hence with $\Delta_0 > 0$ the hypotheses for testing superiority are as follows:

$$H_0 : p_1 - p_2 \leq \Delta_0 \quad \text{versus} \quad H_1 : p_1 - p_2 > \Delta_0 \tag{4.9}$$

or

$$H_0 : \Delta \leq \Delta_0 \quad \text{versus} \quad H_1 : \Delta > \Delta_0 \tag{4.10}$$

In clinical trials, other than testing for non-inferiority and superiority, testing for equivalence is also very common. Test for equivalence is used to demonstrate the similarities or lack of differences between two drugs against a prespecified reference margin Δ_0. The hypotheses testing problem for equivalence can be stated as follows:

$$H_0 : |p_1 - p_2| \geq \Delta_0 \quad \text{versus} \quad H_1 : |p_1 - p_2| < \Delta_0 \tag{4.11}$$

or, equivalently,

$$H_0 : |\Delta| \geq \Delta_0 \quad \text{versus} \quad H_1 : |\Delta| < \Delta_0. \tag{4.12}$$

The rejection of the null hypotheses in the above leads to the acceptance of the hypotheses of equivalence of the two drugs. In order to carry out the test for equivalence, it is useful to consider the following two one-sided hypotheses testing (TOST) problems (see Chow, Shao, and Wang [25] for further details):

$$H_{0a} : \Delta \leq -\Delta_0 \quad \text{versus} \quad H_{1a} : \Delta > -\Delta_0 \tag{4.13}$$

$$H_{0b} : \Delta \geq \Delta_0 \quad \text{versus} \quad H_{1b} : \Delta < \Delta_0 \tag{4.14}$$

Notice that the null hypotheses H_0 specify the set of values of Δ which is the union of the sets of values specified under H_{0a} and H_{0b}, often written as $H_0 = H_{0a} \cup H_{0b}$. Similarly, the alternative hypotheses H_1 specify the set of values of Δ which is the intersection of the sets of values specified under H_{1a} and H_{1b}, often written as $H_1 = H_{1a} \cap H_{1b}$. Later in this chapter, we use this formulation to implement the test for equivalence of two drugs using data from clinical trials.

4.2.2 Asymptotic methods

In this section, we introduce some of the commonly used asymptotic methods for the construction of confidence intervals for the difference between two binomial proportions.

4.2.2.1 Wald interval

Let p_i be the probability of success (also known as risks) of population (i), $i = 1, 2$, and $\Delta = p_1 - p_2$ be the risk difference. Then for the observed data as denoted in Table 4.1, the estimates of the risk difference $\widehat{\Delta}$ is $\widehat{p}_1 - \widehat{p}_2$ with $\widehat{p}_i = x_i/n_i$, and the standard error (SE) is given by

$$SE(\widehat{p}_1 - \widehat{p}_2) = \sqrt{\frac{\widehat{p}_1(1-\widehat{p}_1)}{n_1} + \frac{\widehat{p}_2(1-\widehat{p}_2)}{n_2}}. \tag{4.15}$$

Hence, an approximate $100(1-\alpha)\%$ confidence interval for $\Delta = p_1 - p_2$ is given by:

$$(\widehat{p}_1 - \widehat{p}_2) \pm z_{\alpha/2}\sqrt{\frac{\widehat{p}_1(1-\widehat{p}_1)}{n_1} + \frac{\widehat{p}_2(1-\widehat{p}_2)}{n_2}}, \tag{4.16}$$

where $z_{\alpha/2}$ is the $100(1-\alpha/2)$-th percentile of the standard normal distribution. The above interval is known as the Wald confidence interval.

This confidence interval is widely used in practice and the method is readily available in most software packages. However, this interval suffers from serious erratic behavior of its coverage probabilities as shown in Figure 4.1. Thus, the reliability of the interval estimate, especially for small and moderate sample sizes, is questionable.

The continuity corrected Wald interval due to Yates ([101]) is given by

$$(\widehat{p}_1 - \widehat{p}_2) \pm \left[z_{\alpha/2}SE(\widehat{p}_1 - \widehat{p}_2) + \frac{1}{2}\left(\frac{1}{n_1} + \frac{1}{n_2}\right)\right]. \tag{4.17}$$

The coverage probabilities of the continuity corrected interval are shown in Figure 4.2. The coverage probabilities of the continuity corrected interval have improved for most of the values of p_1, except near the boundaries of the interval $[0, 1]$.

Illustration: Wald method and continuity corrected Wald method

We will demonstrate these methods using the data of pivotal TAXUS ATLAS trial [92]. This was a double-blinded randomized clinical trial designed to compare the performance of the next-generation TAXUS Liberte (drug-eluting stent) to that of TAXUS Express (a bare-metal stent). This was a

FIGURE 4.1
Coverage probabilities of Wald confidence intervals for $n_1 = 25$ and $n_2 = 25$.

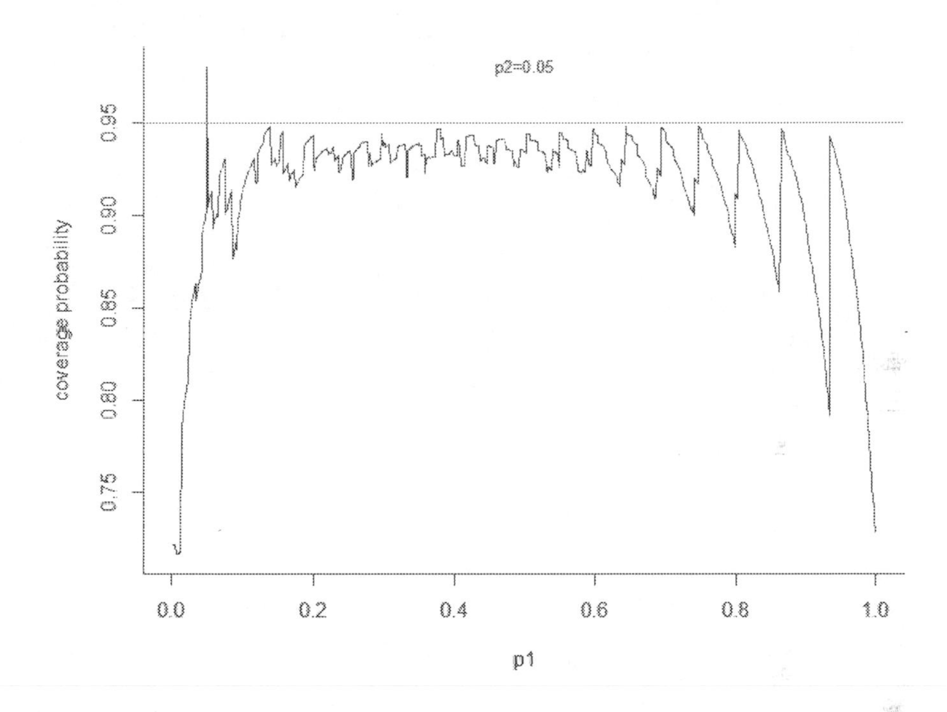

non-inferiority trial with a non-inferiority margin of 3% at $\alpha = 0.05$. The primary endpoint was "Target Vessel Revascularization at 9 months", say, TVR. The trial results are shown in Table 4.2.

For testing non-inferiority of TAXUS Liberte, the statistical hypotheses to be tested are:

$$H_0 : p_1 - p_2(\Delta) \leq -.03(\Delta_0) \quad H_1 : p_1 - p_2(\Delta) > -0.03(\Delta_0), \qquad (4.18)$$

where p_1 and p_2 are the success probabilities of TAXUS Liberte and TAXUS Express, respectively.
In such a setting, the non-inferiority(NI) margin is specified when the trial is designed, and the statistical inference is carried out using the confidence interval approach. This approach is commonly used in clinical trials as it is also recommended in the FDA's and the EMEA's guidelines for testing non-inferiority ([66], [37]). The confidence interval boundary is usually compared to the NI margin to evaluate if non-inferiority is achieved. Using a statistical software, one can compute a 90% confidence interval of Δ (since $\alpha = 0.05$)

FIGURE 4.2
Coverage probabilities of Wald (corrected) confidence intervals for $n_1 = 25$ and $n_2 = 25$.

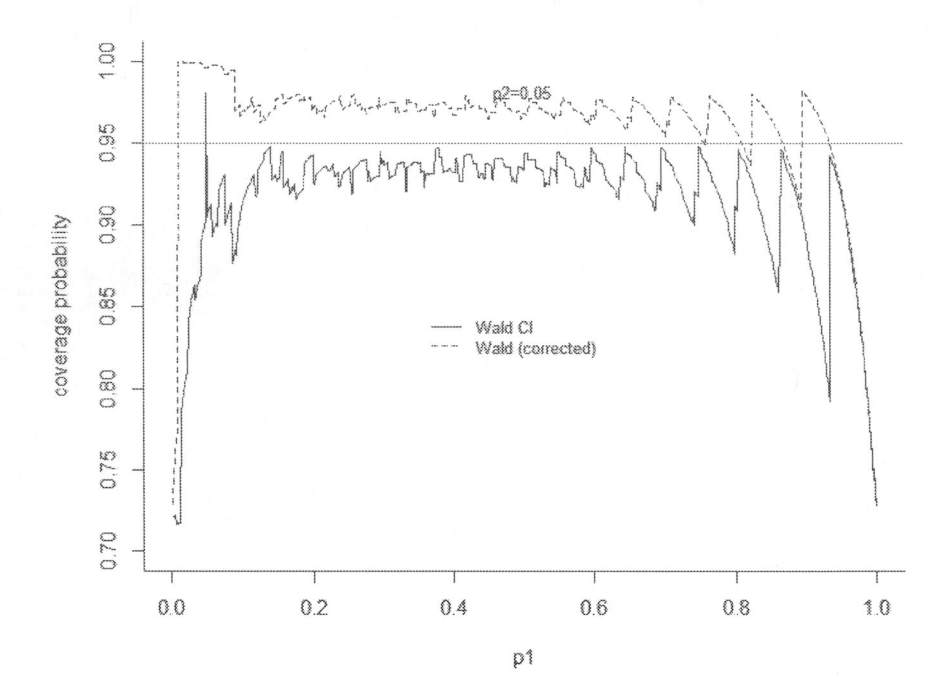

and test for the non-inferiority. The following is the SAS code for computing a 90% confidence interval:

```
data DESAtlas;
   input Exposure Response Count;
   datalines;
   1 1  68
   1 0 787
   2 1  67
   2 0  889
;
proc freq data=DESAtlas;
tables Exposure*Response/riskdiff (noninf margin=0.03) alpha=0.05;
weight Count;
run;
```

In the above SAS *proc freq* code, *tables Exposure*Response* is used to specify the input table with *Exposure* as the population variable indicating the populations as either 1 or 2, and *Response* as the response variable. The

TABLE 4.2
TAXUS ATLAS trial results at 9 months

	TAXUS Express	TAXUS Liberte'
9-month TVR	7.01% (67/956)	7.95% (68/855)

option *weight count* provides the cell counts of each cell of the 2×2 table. The option *riskdiff* (Δ_0 *=0.03)* invokes the risk difference option ($p_1 - p_2 = \Delta$) for testing non-inferiority with a margin = 0.03. Finally, the syntax *alpha=0.05* is for specifying testing 1-sided non-inferiority at $\alpha = 0.05$.

The SAS output for the example data is shown below:

```
Noninferiority Analysis for the Proportion (Risk) Difference

 H0: P1 - P2 <= -Margin     Ha: P1 - P2 > -Margin

             Margin = 0.03     Wald Method

 Proportion Difference    ASE (Sample)          Z     Pr > Z

         -0.0094     0.0124       1.6572     0.0487

 Noninferiority Limit     90% Confidence Limits

               -0.0300          -0.0298      0.0109
```

Note that in the above SAS output, the lower limit of 90% is −0.0298, which is marginally missing the NI margin of 3%. As a result, one has to reject the null hypothesis in favor of the alternative hypothesis. In the Wald method, because of the duality between the confidence interval and hypothesis testing (see Section 1.2.2), the test result using the z-statistic directly corresponds to the inference using the confidence interval. In the edited SAS output above, we see that the p-value is 0.0487, which is marginally lower than the significance level alpha = 0.05, indicating that the lower confidence limit has marginally missed the target NI margin.

We next carry out the same test using the continuity corrected interval. The SAS code is given below:

```
proc freq data=DESAtlas;
tables Exposure*Response/riskdiff (correct noninf margin=0.03)
                         alpha=0.05;
weight Count;
run;
```

In the above PROC FREQ code *correct* is specified to execute the Wald test using the continuity correction method. The SAS output is given below.

```
Noninferiority Analysis for the Proportion (Risk) Difference

          H0: P1 - P2 <= -Margin    Ha: P1 - P2 > -Margin

                    Margin = 0.03    Wald Method

          Proportion Difference    ASE (Sample)          Z     Pr > Z

                      -0.0094          0.0124     1.5679    0.0585

 Noninferiority Limit    90% Confidence Limits

             -0.0300          -0.0310      0.0121
```

Note that this confidence interval is a little wider, and consequently, it fails to reject the null hypothesis ($p = 0.0585$).

4.2.2.2 Agresti and Caffo interval

Wald interval performs poorly when the sample size is small, or even moderate (see [15], [1]). Agresti and Caffo [1] propose a simple adjustment to the Wald method. For each binomial population they propose adding one success and one failure to the observed data. Therefore, based on the Agresti and Caffo's formulation, the estimator of the risk difference is $\widehat{\Delta} = \widetilde{p}_1 - \widetilde{p}_2$, where $\widetilde{p}_i = \dfrac{(x_i+1)}{(n_i+2)}, i = 1, 2$. Following equation 4.15, the estimated standard error of $\widehat{\Delta}$ is given by

$$SE(\widehat{\Delta}) \quad = \quad \sqrt{\frac{\widetilde{p}_1(1-\widetilde{p}_1)}{n_1} + \frac{\widetilde{p}_2(1-\widetilde{p}_2)}{n_2}}, \tag{4.19}$$

and thus, the $100(1-\alpha)\%$ confidence interval due to Agresti and Caffo is computed as follows:

$$(\widetilde{p}_1 - \widetilde{p}_2) \pm z_{\alpha/2}\sqrt{\frac{\widetilde{p}_1(1-\widetilde{p}_1)}{n_1} + \frac{\widetilde{p}_2(1-\widetilde{p}_2)}{n_2}}. \tag{4.20}$$

Agresti and Caffo's (**AC**) method can be easily implemented by writing a simple code in standard software. Also standard statistical software compute this interval. In particular, Agresti has made available `R` code implementing several methods of computing the confidence interval of risk difference on his website (`http://www.stat.ufl.edu/~aa/cda/R/`). One can also use the R-package *PropCIs* to compute a confidence interval using this method.
Compared to the Wald interval, the Agresti and Caffo (**AC**) interval has a better coverage property. Notice that, in Figure 4.3, for fixed $p_2 = 0.05$, there are values of p_1 around 0.5, for which the method is unable to guarantee the nominal coverage probability.
In the previous chapter, we have discussed the application of confidence interval in testing for non-inferiority. For the non-inferiority trial, because of the

FIGURE 4.3
Coverage probabilities ofAgresti and Caffo confidence intervals for $n_1 = 25$ and $n_2 = 25$.

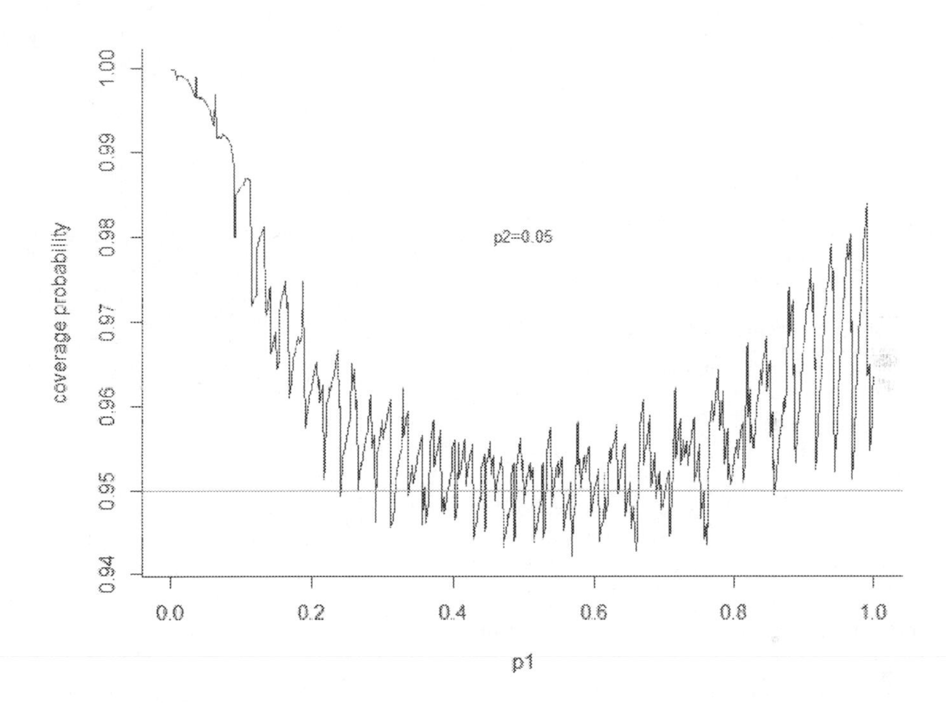

form of statistical hypothesis, a one-sided confidence interval with confidence level $1 - \alpha = 0.95$ is required. Thus, one-sided 95% confidence intervals are derived from two-sided 90% confidence intervals. However, in the following, we present an example with a two-sided 95% confidence interval, hence, leading to a 97.5% one-sided interval, and a test for non-inferiority at level 2.5%.

Illustration: Agresti and Caffo (AC) interval

We illustrate the application of **AC** interval using the data set from GEMINI 3 trial. This trial is a double-blinded randomized clinical trial to evaluate the efficacy of Vedolizumab to placebo in the treatment of Crohn's Disease, a type of Inflammatory Bowel Disease (IBD). The claim is, Vedolizumab is superior to placebo. One of the secondary endpoints in this trial is remission at week 6. The observed remission rate at week 6 for the intention to treat population (ITT) is 25/207(12.1%) in the placebo arm and 40/209(19.1%) in the Vedolizumab arm. In order to compare Vedolizumab with placebo, the

95% confidence interval of the placebo corrected delta (i.e., $p_2 - p_1$) can be computed using the following SAS code:

```
data CD6wkRemisson;
input Exposure Response Count;
datalines;
1 1 25
1 0 182
2 1 40
2 0 169
;
proc freq data=CD6wkRemisson;
tables Exposure*Response/riskdiff(CL=AC) alpha=0.05;
weight Count;
run;
```

The SAS output is the following:

```
     Statistics for Table of Exposure by Response

  Confidence Limits for the Proportion (Risk) Difference
          Column 1 (Response = 0)
       Proportion Difference = 0.0706

  Type              95% Confidence Limits

  Agresti-Caffo      0.0003         0.1396
```

In the PROC FREQ syntax, the option riskdiff(CL=AC) invokes the risk difference option using Agresti and Caffo interval. In the SAS output the 95% confidence interval is shown as $(0.0003, 0.1396)$. Since the lower limit of 95% confidence interval is very close to 0 but does not include 0, we can conclude that these two arms are marginally different.

One can also run the following code with R package *PropCIs* to compute the confidence interval due to Agresti and Caffo:

```
library(PropCIs)
wald2ci(40, 209,25, 207, 0.95, adjust="AC")
```

The following output is shown in R:

```
95 percent confidence interval:
 0.0002522675 0.1395694971
sample estimates:
[1] 0.06991088
```

Note that the output from R looks slightly different than the one produced by SAS. It is because the SAS output is rounded up to 4 decimal places; hence, for all practical purposes, they are essentially the same.

4.2.2.3 Newcombe's score interval

As stated in the previous chapter, in the context of a single binomial confidence interval, the Wilson score interval for population i, $i = 1, 2$, can be computed using the following equations:

$$\left(\widehat{p}_i + z_{\alpha/2}^2/2n_i\right) \pm \left(z_{\alpha/2}\sqrt{(\widehat{p}_i(1-\widehat{p}_i) + z_{\alpha/2}^2})/(1 + z_{\alpha/2}^2/n_i)\right) \quad (4.21)$$

Let (l_i, u_i) be the score interval for $p_i, i = 1, 2$; then for the difference of two independent binomial proportions Δ, Newcombe ([60]) proposes an interval $(\underline{\Delta}_{NW}, \overline{\Delta}^{NW})$ where,

$$\underline{\Delta}_{NW} = (\widehat{p}_1 - \widehat{p}_2) - z_{\alpha/2}\sqrt{l_1(1-l_1)/n_1 + u_2(1-u_2)/n_2}, \quad (4.22)$$

and

$$\overline{\Delta}^{NW} = (\widehat{p}_1 - \widehat{p}_2) + z_{\alpha/2}\sqrt{u_1(1-u_1)/n_1 + l_2(1-l_2)/n_2}. \quad (4.23)$$

FIGURE 4.4
Coverage probabilities of Newcombe confidence intervals for $n_1 = 25$ and $n_2 = 25$.

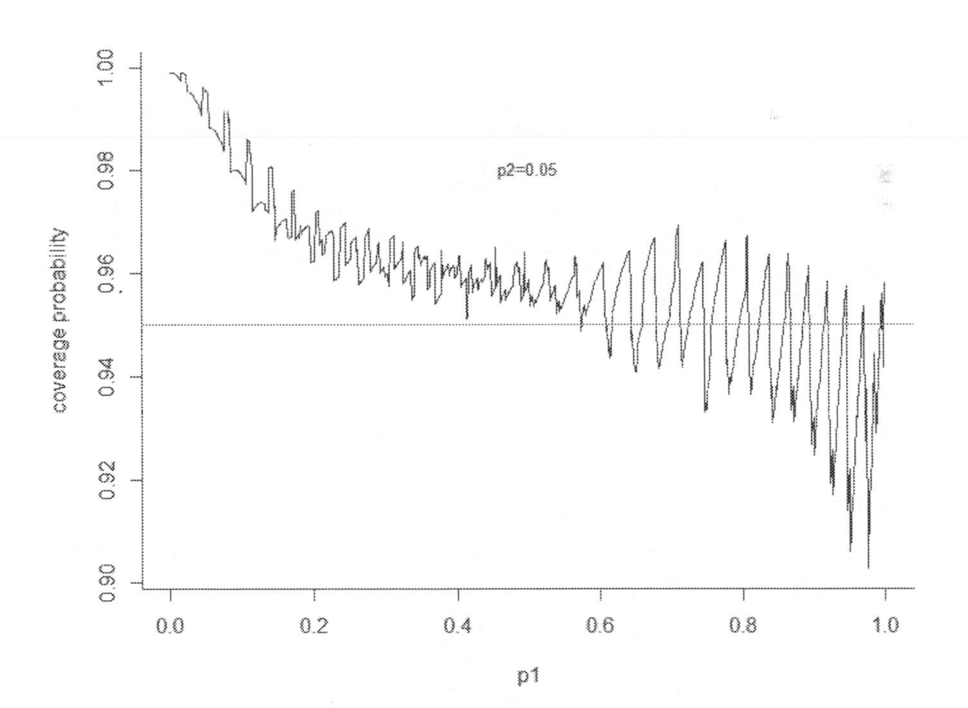

The Newcombe method, also known as the score method, is known to have good coverage probabilities. For sample size $n_1 = 25$ and $n_2 = 25$, Figure 4.4 shows the coverage probabilities for fixed $p_2 = 0.05$, and varying p_1. It appears that the coverage probabilities fail to achieve the nominal level for p_1 more than 0.5.

FIGURE 4.5
Coverage probabilities of Wilson (corrected) confidence intervals for $n_1 = 25$ and $n_2 = 25$.

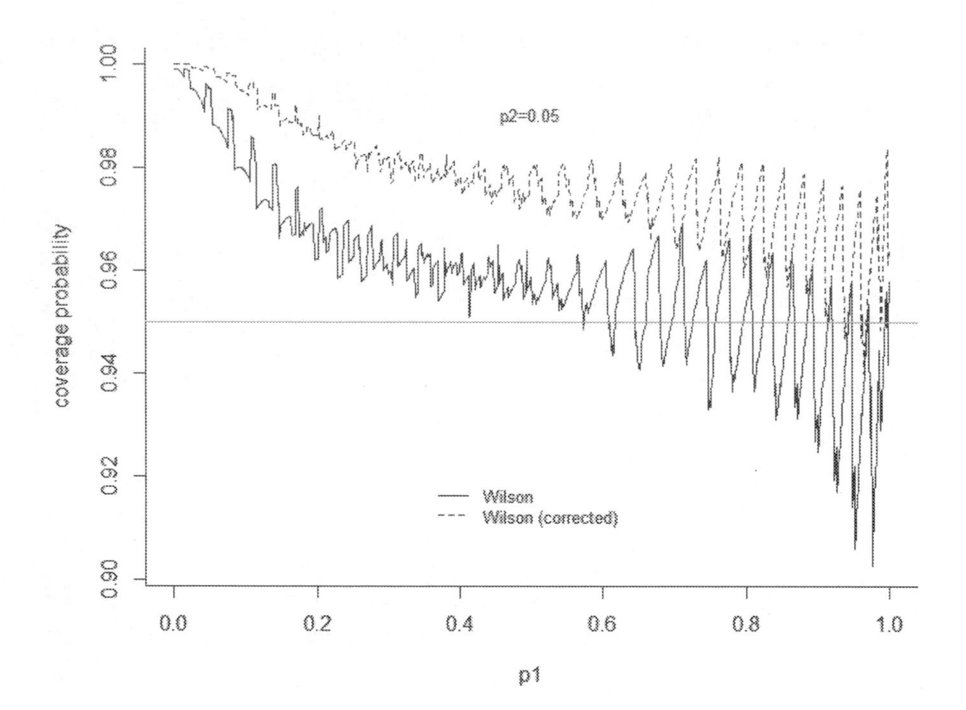

Illustration: Newcombe's score interval

The above method is illustrated using data from the Atlas trial used in Section 4.2.2.1. The non-inferiority hypothesis in this setting can be tested by running the following SAS code:

```
data DESAtlas;
   input Exposure Response Count;
   datalines;
   1 1  68
```

```
   1 0 787
   2 1  67
   2 0  889
;

proc freq data=DESAtlas;
tables Exposure*Response/riskdiff (noninf margin=0.03 method=NEWCOMBE)
                         alpha=0.05;
weight Count;
run;
```

In the above code, *method= NEWCOMBE* option in PROC FREQ invokes the Newcombe's method, and the following SAS output is related to the non-inferiority test.

```
Noninferiority Analysis for the Proportion (Risk) Difference

 H0: P1 - P2 <= -Margin     Ha: P1 - P2 > -Margin

          Margin = 0.03    Newcombe Score Method

 Proportion Difference    Noninferiority Limit    90% Confidence Limits

           -0.0094                  -0.0300         -0.0302      0.0109
```

Since the lower limit of the confidence interval fails to cross the specified margin of −0.300, the test fails to reject the null hypothesis or, equivalently fails to accept the hypotheses of non-inferiority at 5% level of significance.

Like Wald interval, Newcombe ([60]) also proposes a correction factor to the confidence limits. For sample size $n_1 = 25$ and $n_2 = 25$, Figure 4.5 shows the coverage probabilities for fixed $p_2 = 0.05$, and varying p_1.

Using the following SAS code, for Atlas trial data, the non-inferiority hypothesis can be tested with Newcombe (corrected) interval as follows:

```
data DESAtlas;
   input Exposure Response Count;
   datalines;
   1 1  68
   1 0 787
   2 1  67
   2 0  889
;
proc freq data=DESAtlas;
tables Exposure*Response/riskdiff (correct noninf margin=0.03
                         method=NEWCOMBE) alpha=0.05;
weight Count;
run;
```

In the above code, correct option used with PROC FREQ is specified to run method=NEWCOMBE with continuity correction, resulting in the following SAS output:

```
Noninferiority Analysis for the Proportion (Risk) Difference

   H0: P1 - P2 <= -Margin     Ha: P1 - P2 > -Margin

        Margin = 0.03    Newcombe Method

Proportion Difference    Noninferiority Limit    90% Confidence Limits

         -0.0094                -0.0300              -0.0309      0.0117

The confidence limits include a continuity correction.
```

Although the lower limit -0.0309 of the corrected interval is slightly different from the lower limit -0.031 of the uncorrected interval, the inference remains the same.

4.2.2.4 Profile likelihood based interval

For the observed frequency table shown in Table 4.1, the likelihood can be written as

$$L(p_1, p_2 : \boldsymbol{x}) = \prod_{j=1}^{2} \binom{n_j}{x_j} p_j^{x_j} (1 - p_j)^{n_j - x_j},$$

where the symbols have their usual significance. The likelihood can be equivalently written in terms of Δ, the parameter of interest, and p_1, a nuisance parameter as shown below:

$$L(p_1, \Delta : \boldsymbol{x}) = \prod_{j=1}^{2} \binom{n_j}{x_j} p_1^{x_1} (1-p_1)^{n_1-x_1} (p_1 - \Delta)^{x_2} (1 - p_1 + \Delta)^{n_2 - x_2}. \quad (4.24)$$

Now, similar to Agresti and Caffo [1]'s adjustment by adding a non-negative number a to each cell of the observed table, the kernel of the observed log-likelihood becomes

$$\begin{aligned} l(p_1, \Delta) &= (x_1 + a)ln(p_1) + (n_1 - x_1 + a)ln(1 - p_1) + (x_2 + a)ln(p_1 - \Delta) \\ &\quad +(n_2 - x_2 + a)ln(1 - p_1 + \Delta). \end{aligned} \quad (4.25)$$

By inverting the likelihood ratio test statistic, Venzon and Moolgavkar [94] propose an efficient method of computing profile likelihood-based confidence interval. Pradhan and Banerjee [68] show that for the choice $a = 0.5$ in equation (4.24), the profile likelihood-based confidence interval leads to intervals with better coverage probabilities and shorter expected lengths. The approximate $1 - \alpha$ profile likelihood-based interval for Δ is

$$\{\Delta : 2[l(\widehat{\Delta}, \widehat{p}_1) - l(\Delta, \widetilde{p}_1)] \leq \chi_1^2(\alpha)\}, \quad (4.26)$$

where $\widehat{\Delta}$ and $\widehat{p}_1$ are the unrestricted maximum likelihood estimates of Δ and p_1 respectively, $\widetilde{p}_1$ is the restricted maximum likelihood of p_1 for given Δ, and $\chi_1^2(\alpha)$ is the upper 100α percentile point of the chi-square distribution. Following Venzon and Moolgavkar [94], the confidence interval for Δ, is the admissible solution to the following system of non-linear equations:

$$\begin{bmatrix} l(\widehat{\Delta}, \widehat{p}_1) - l(\Delta, p_1) - \frac{1}{2}\chi_1^2(\alpha) \\ \frac{\partial l(\Delta, p_1)}{\partial p_1} \end{bmatrix} = 0.$$

The SAS/IML implementation of the method based on Pradhan and Banerjee [68] is given below.

```
%macro p1p2_profile(sample1= ,sample2= ,success1= ,success2= , alpha= ,
 a= );
%if &success2=0 %then %do;
%let s2=%eval(&success1);
%let s1=%eval(&success2);
%let n1=%eval(&sample2);
%let n2=%eval(&sample1);
%end;
%else %do;
%let s1=%eval(&success1);
%let s2=%eval(&success2);
%let n2=%eval(&sample2);
%let n1=%eval(&sample1);
%end;
proc iml;
Start likelihood(x)global(n1,n2,s1,s2) ;
      sum1=0.0;sum2=0.0;sum3=0.0;sum4=0.0;
      if (x[2]>0)&(x[1]>=x[2]) then do;
      sum1=(s1+&a )#log(x[1]);

      dem1=1.0-x[1];
      if dem1<=0 then dem1=1e-6;
      sum2=(n1-s1+&a )#log(dem1);

      demo2=x[1]-x[2];
      if demo2 <=0 then demo2=1e-6;
      sum3=(s2+&a )#log(demo2);

      dem2=1.0-x[1]+x[2];
      if dem2<=0 then dem2=1e-6;
      sum4=(n2-s2+&a )#log(dem2);

      f=sum1+sum2+sum3+sum4; end;
      else do;
      sum1=(s1+&a )#log(x[1]);

      dem3=1.0-x[1];
      if dem3<=0 then dem3=1e-6;
      sum2=(n1-s1+&a )#log(dem3);

      dem4=x[1]-x[2];
      if dem4<=0 then dem4=1e-6;
      sum3=(s2+&a )#log(dem4);

      dem5=1.0-x[1]+x[2];
      if dem5 <=0 then dem5=1e-6;
      sum4=(n2-s2+&a )#log(dem5);
      f=sum1+sum2+sum3+sum4; end;
      return (f);
finish likelihood;

start Gradient(x)global(n1,n2,s1,s2);
      g=j(1,2,1e-16);
      if (x[2]>=0)&(x[1]>x[2]) then do;
```

```
      g[1]=(s1+&a )/x[1] -((n1-s1+&a )/(1.0-x[1]))+((s2+&a )/(x[1]-x[2]))
           -((n2-s2+&a )/(1.0-x[1]+x[2]));
      g[2]=-((s2+&a )/(x[1]-x[2]))+((n2-s2+&a )/(1.0-x[1]+x[2]));
      end;
      else do;
      g[1]=(s1+&a )/x[1] -((n1-s1+&a )/(1.0-x[1]))+((s2+&a )/(x[1]-x[2]))
           -((n2-s2+&a )/(1.0-x[1]+x[2]));
      g[2]=-((s2+&a )/(x[1]-x[2]))+((n2-s2+&a )/(1.0-x[1]+x[2]));
      end;
      return (g);
finish Gradient;

 n = 2; n1=&n1 ;n2=&n2 ;s1=&s1 ;s2=&s2;
 x0 ={1e-6 -0.9999};
 optn = {1 0};
 con = { 1.e-6 -0.99999 . .,
 0.99999 0.99999 . . ,
                 1 -1 1 1e-6,
                 1 -1 -1 0.99999};
 call nlptr(rc,xres,"likelihood",x0,optn,con,,,,"Gradient");
 xopt = xres`; fopt = likelihood(xopt);

       call nlpfdd(f,g,hes2,"likelihood",xopt,,"Gradient");

 start plgrad(x) global(like,ipar,lstar);
      like = likelihood(x);
      grad = Gradient(x);
      grad[ipar]=like-lstar;
      return(grad`);
 finish plgrad;

      prob=&alpha ;
      xlb=j(2,1,-1);
      xub=j(2,1,1);
      /* quantile of chi**2 distribution */
      chqua = cinv(1-prob,1);like=fopt; lstar = fopt - .5 * chqua;

 optn = {2 0};
 do ipar = 1 to 2;
 /* Implementation of Venzon & Moolgavkar (1988)*/
      if ipar=1 then ind = 2; else ind = 1;
       delt = - inv(hes2[ind,ind]) * hes2[ind,ipar];
      alfa = - (hes2[ipar,ipar] - delt` * hes2[ind,ipar]);
       if alfa > 0 then alfa = .5 * sqrt(chqua / alfa);
       else do;
         print "Bad alpha";
         alfa = .1 * xopt[ipar];
      end;
      if ipar=1 then delt = 1 || delt;
       else delt = delt || 1;
       x0 = xopt + (alfa * delt)`;
       con2 = con; con2[1,ipar] = xopt[ipar];
        tc={2000,5000};
       call nlplm(rc,xres,"plgrad",x0,optn,con2,tc );

       f = plgrad(xres); s = ssq(f);

        if (s <1.e-6) then xub[ipar] = xres[ipar];
        else xub[ipar] =1.0;
        x0 = xopt - (alfa * delt)`;
       con2[1,ipar] = con[1,ipar]; con2[2,ipar] = xopt[ipar];
        tc={2000,5000};
       call nlplm(rc,xres,"plgrad",x0,optn,con2,tc);
       f = plgrad(xres); s = ssq(f);
       if (s < 1.e-6) then xlb[ipar] = xres[ipar];
       else xlb[ipar] = -1.0;
```

```
 end;

if &success2=0 then do;
ci=-xub||-xlb;
end;
else do;
ci=xlb||xub; end;
create Profile from ci[colname={'Lower','Upper'} ] ;
append from ci;
close Profile;
run;
quit;
run;
data Profile; set Profile;
if _n_=1 then delete;
run;

Proc print data=Profile noobs label;
title1 "************************************************************";
title2 "Confidence interval on difference of two binomial proportions ";
title3 "Based on Weighted Profile Likelihood Method                   ";
title4 "Proportions are P1=&success1/&sample1 and P2=&success2/&sample2";
title5 "************************************************************";
run;
%mend;
```

The above SAS/IML code usually produces a sensible confidence interval, but it may produce erroneous confidence limits in some rare instances. In such cases, SAS's log window shows WARNING messages. Using the above code, one can easily get a confidence interval by using the rules stated in Equations 4.2 and 4.3.

For $n_1 = 25$, $n_2 = 25$, and $p_2 = 0.05$ the coverage distribution is given Figure 4.6. In this set up, in general, the profile likelihood method shows good coverage although there are few points p_1 greater than 0.75 for which nominal coverage probabilities are not attained, and the lowest attained coverage probability is around 90%.

Illustration: Profile likelihood method

Consider the Atlas trial data introduced earlier, we want to compute the 90% confidence interval with $a = 0.5$ using the profile likelihood method. For this purpose, the above SAS macro can be used using the following modification:

```
%p1p2_profile(sample1=956, sample2=855, success1=67, success2=68,
alpha=0.1, a=0.5);
```

This SAS macro shows the following output:

```
      Lower        Upper

   -0.030138    0.010893
```

The output shows that the lower limit barely misses the non-inferiority margin 3% and hence fails to reject the null hypothesis.

FIGURE 4.6
Coverage probabilities of Profile likelihood confidence intervals for $n_1 = 25$ and $n_2 = 25$.

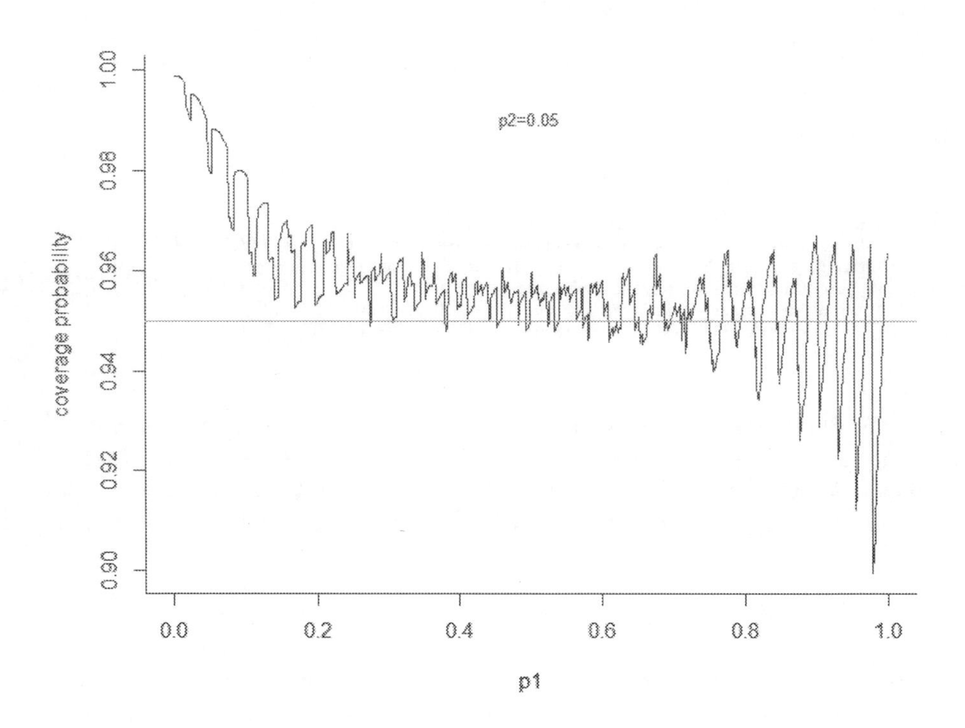

4.2.2.5 Farrington and Manning (score) interval

The score statistic due to Farrington and Manning [35] (FM) for testing H_0: $\Delta = \Delta_0$ versus H_1: $\Delta \neq \Delta_0$ is given by

$$S(\boldsymbol{x}) = \frac{\widehat{p}_1 - \widehat{p}_2 - \Delta_0}{\sqrt{\frac{(\widetilde{p}_1)(1-\widetilde{p}_1)}{n_1} + \frac{(\widetilde{p}_2)(1-\widetilde{p}_2)}{n_2}}}.$$

Here $\widehat{p}_j = x_j/n_j$ for $j = 1, 2$, and $\widetilde{p}_1$ and $\widetilde{p}_2$ are the maximum likelihood estimates of p_1 and p_2, respectively, under the restriction that $p_1 - p_2 = \Delta_0$. The test statistic must be defined separately if $x_1 = x_2 = 0$ and $\Delta_0 = 0$. However, this configuration does not occur when the score statistic is inverted for computing confidence intervals for Δ. Farrington and Manning derived the restricted maximum likelihood estimates $\widetilde{p}_1$ and $\widetilde{p}_2$ by solving the third degree likelihood equation

$$\sum_{k=1}^{3} L_k p_1^k = 0 \quad \text{for } p_1 \in [\max\{0, -\Delta_0\}, \min\{1, 1-\Delta_0\}] \ ,$$

where

$$\begin{aligned} L_3 &= N, \\ L_2 &= (n_1 + 2n_2)\Delta_0 - N - x_1 - x_2, \\ L_1 &= (n_2\Delta - N - 2x_2)\Delta_0 + x_1 + x_2, \\ L_0 &= x_2\Delta_0(1-\Delta_0), \end{aligned}$$

and setting $\widetilde{p}_1 = \widetilde{p}_2 + \Delta_0$; here $N = n_1 + n_2$. Under H_0, the test statistic $S(\boldsymbol{x})$ has mean 0 and variance 1.

Solving these equations we can show the values of $\widetilde{p}_1$ and $\widetilde{p}_2$ as follows:

$$\begin{aligned} \widetilde{p}_2 &= 2p\cos a - \frac{L_2}{3L_3} \\ \widetilde{p}_1 &= \widehat{p}_2 + \Delta_0 \end{aligned}$$

where

$$\begin{aligned} q &= L_2^3/(3L_3)^3 - L_1L_2/6L_3^2 + L_0/2L_3 \\ p &= sign(q)\sqrt{L_2^2/(3L_3)^2 - L_1/3L_3} \\ a &= \frac{1}{3}[\pi + cos^{-1}(q/p^3)] \end{aligned}$$

The asymptotic $100(1-\alpha)\%$ confidence interval due to **FM** is obtained by inverting one-sided tests based on $S(\boldsymbol{x})$; this leads to the interval endpoints defined as

$$1 - \Phi\left(\frac{\widehat{p}_1 - \widehat{p}_2 - \underline{\Delta}^{FM}}{\sqrt{\left(\frac{(\widetilde{p}_1)(1-\widetilde{p}_1)}{n_1} + \frac{(\widetilde{p}_2)(1-\widetilde{p}_2)}{n_2}\right)}}\right) = \frac{\alpha}{2} = \Phi\left(\frac{\widehat{p}_1 - \widehat{p}_2 - \overline{\Delta}^{FM}}{\sqrt{\left(\frac{(\widetilde{p}_1)(1-\widetilde{p}_1)}{n_1} + \frac{(\widetilde{p}_2)(1-\widetilde{p}_2)}{n_2}\right)}}\right)$$

Due to the increasing popularity of the score method, the **FM** interval has been implemented in the most commonly available software, including R. Newcombe ([60]) compared several methods for computing confidence intervals for the difference between two binomial proportions and recommended the FM method, especially for small-sample sizes. Figure 4.7 shows the coverage probability distribution of **FM** interval.

Under this setup, with $p_2 = 0.05$, it can be seen that there are several points around $p_1 < 0.3$ that have under coverage. A more comprehensive coverage comparison for all points (p_1, p_2) is discussed later in this chapter.

FIGURE 4.7
Coverage probabilities of Farrington and Manning confidence intervals for $n_1 = 25$ and $n_2 = 25$.

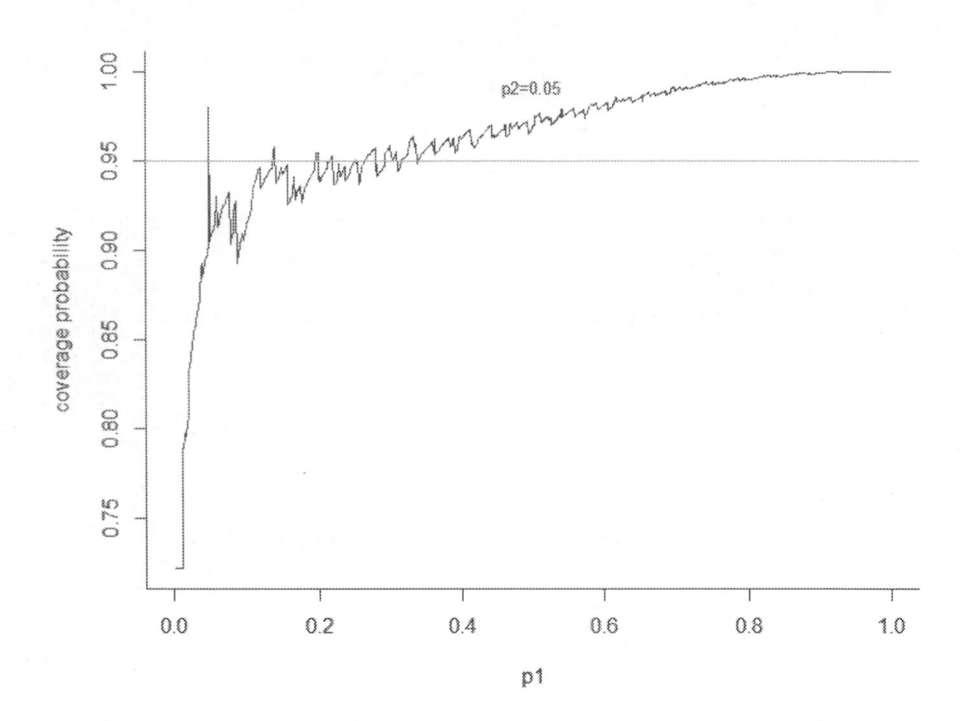

Illustration: Farrington and Manning (score) interval

To illustrate this interval, let us consider the Atlas trial data and compute the 90% confidence interval using SAS. The following SAS code can be used:

```
data DESAtlas;
   input Exposure Response Count;
   datalines;
   1 1  68
   1 0 787
   2 1  67
   2 0  889
;
proc freq data=DESAtlas;
tables Exposure*Response/riskdiff (CL=FM) alpha=0.1;
weight Count;
run;
```

The SAS's PROC FREQ shows the following output:

```
Confidence Limits for the Proportion (Risk) Difference
              Column 1 (Response = 0)
    Proportion Difference = -0.0094

   Type                   90% Confidence Limits

   Farrington-Manning     -0.0298        0.0109
```

Since the lower limit of the 90% confidence interval barely excludes -0.03, the null hypothesis is rejected, and thus, the hypothesis of non-inferiority is accepted.

4.2.2.6 Miettinen and Nurminen (score) interval

Miettinen and Nurminen [57] (**MN**) propose a score interval which is almost identical to that of the FM interval. The only difference is, in the denominator of $S(x)$, a correction factor $\frac{N}{N-1}$ is multiplied by the variance.

Among the asymptotic methods, Miettinen and Nurminen (score) interval is known to have good coverage property. For $n_1 = 25$, $n_2 = 25$, and $p_2 = 0.05$ the coverage distribution is given Figure 4.8. In this setup, **MN** interval generally shows good coverage property. However, there are several points where nominal coverage probabilities are not attained, and the lowest coverage probability is around 93%.

Illustration: Miettinen and Nurminen method (score)

To demonstrate this method, we use again the Atlas trial data to compute the 90% confidence interval. The SAS code given below can be used to calculate the confidence intervals using the **MN** method:

```
data DESAtlas;
   input Exposure Response Count;
   datalines;
   1 1  68
   1 0 787
   2 1  67
   2 0  889
;
proc freq data=DESAtlas;
tables Exposure*Response/riskdiff (CL=MN) alpha=0.1;
weight Count;
run;
```

The SAS's PROC FREQ leads to the following output:

```
Confidence Limits for the Proportion (Risk) Difference
              Column 1 (Response = 0)
    Proportion Difference = -0.0094

   Type                   90% Confidence Limits

   Miettinen-Nurminen     -0.0302        0.0109
```

FIGURE 4.8
Coverage probabilities of Miettinen and Nurminen confidence intervals for $n_1 = 25$ and $n_2 = 25$.

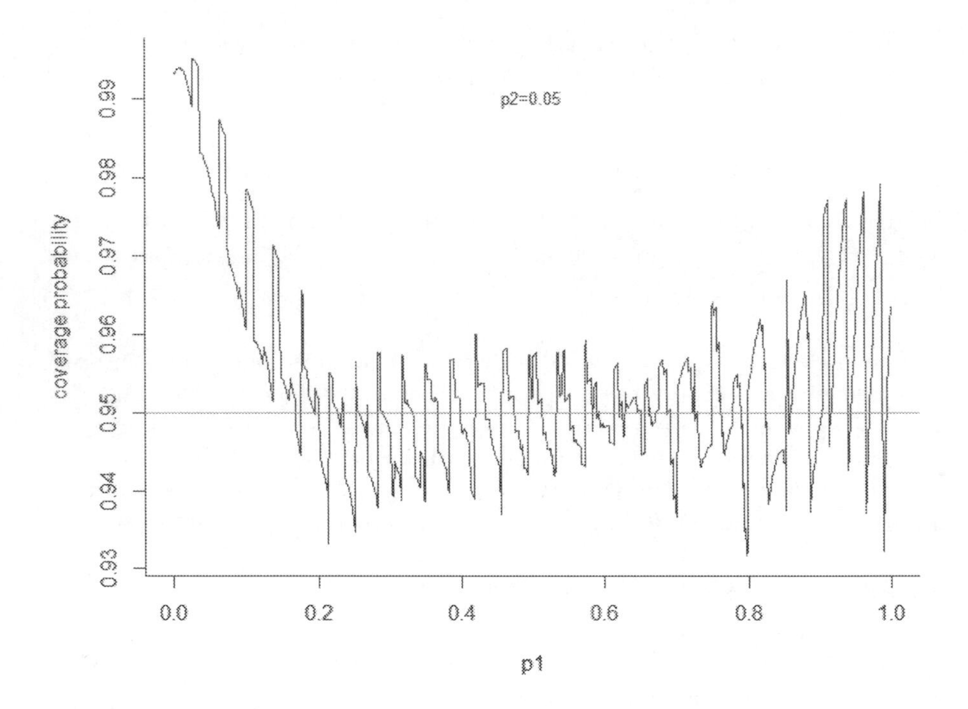

The lower limit of the 90% confidence interval includes 0.03 consistent with other confidence intervals like **AC**. Thus, the null hypothesis cannot be rejected at level $\alpha = 0.05$ with a NI margin 3%. In contrast, the **FM** method fails to include 0.03. To compute both the intervals simultaneously, one can run the following SAS code:

```
data DESAtlas;
   input Exposure Response Count;
   datalines;
   1 1  68
   1 0 787
   2 1  67
   2 0  889
;
proc freq data=DESAtlas;
tables Exposure*Response/riskdiff (CL=(FM MN)) alpha=0.1;
weight Count;
run;
```

In the above SAS code, specifying FM and MN in parenthesis in the option *riskdiff (CL=(FM MN))* invokes both the methods to compute 90% confidence intervals. SAS's PROC FREQ shows the following output:

```
Confidence Limits for the Proportion (Risk) Difference
                 Column 1 (Response = 0)
           Proportion Difference = -0.0094
            Type                    90% Confidence Limits

           Farrington-Manning     -0.0298         0.0109
           Miettinen-Nurminen     -0.0302         0.0109
```

Note that **MN** interval is slightly wider due to the extra multiplicative factor $\frac{N}{N-1}$ in the variance computation.

4.2.2.7 MOVER interval

Newcombe [62] proposes the 'square and add' method for finding confidence interval of Δ from the limits of the confidence intervals of p_1 and p_2. Donner and Zou [32] generalize this approach for finding confidence interval of a linear combination of independent parameters from the limits of the individual confidence intervals, and provide a theoretical justification of this method in terms of local variance estimates recovered. The method is named as MOVER (an acronym for the "Method of Variance Estimate Recovery").
Let the maximum likelihood estimate of p_i be $\widehat{p}_i$ and the corresponding confidence interval is (l_i, u_i) for $i = 1, 2$. Then the MOVER interval of Δ, say, $(\underline{\Delta}, \overline{\Delta})$ can be computed as

$$\begin{aligned} \underline{\Delta} &= (\widehat{p}_1 - \widehat{p}_2) - \sqrt{(\widehat{p}_1 - l_1)^2 + (u_2 - \widehat{p}_2)^2} \\ \overline{\Delta} &= (\widehat{p}_1 - \widehat{p}_2) + \sqrt{(u_1 - \widehat{p}_1)^2 + (\widehat{p}_2 - l_2)^2}. \end{aligned} \tag{4.27}$$

Donner and Zou recommend using Wilson [99] method to compute the intervals $(l_i, u_i), i = 1, 2$. The MOVER interval, for its computation, recovers the uncertainty of the estimate of $p_1 - p_2$ from the limits of the individual confidence intervals of p_1 and p_2.

Figure 4.9 shows the coverage probabilities of **MOVER interval** for varying values of p_1 using Agresti and Coull, Clopper–Pearson, Jeffreys and Wilson Score intervals for $n_1 = 25$, $n_2 = 25$ and $p_2 = 0.05$. In this setup, **MOVER intervals** typically show good coverage properties. However, for some values of p_1 the coverage probabilities are much higher than 95% nominal level, indicating that the **MOVER intervals** are conservative (wider confidence intervals).

FIGURE 4.9
Coverage probabilities of MOVER intervals for $n_1 = 25$ and $n_2 = 25$.

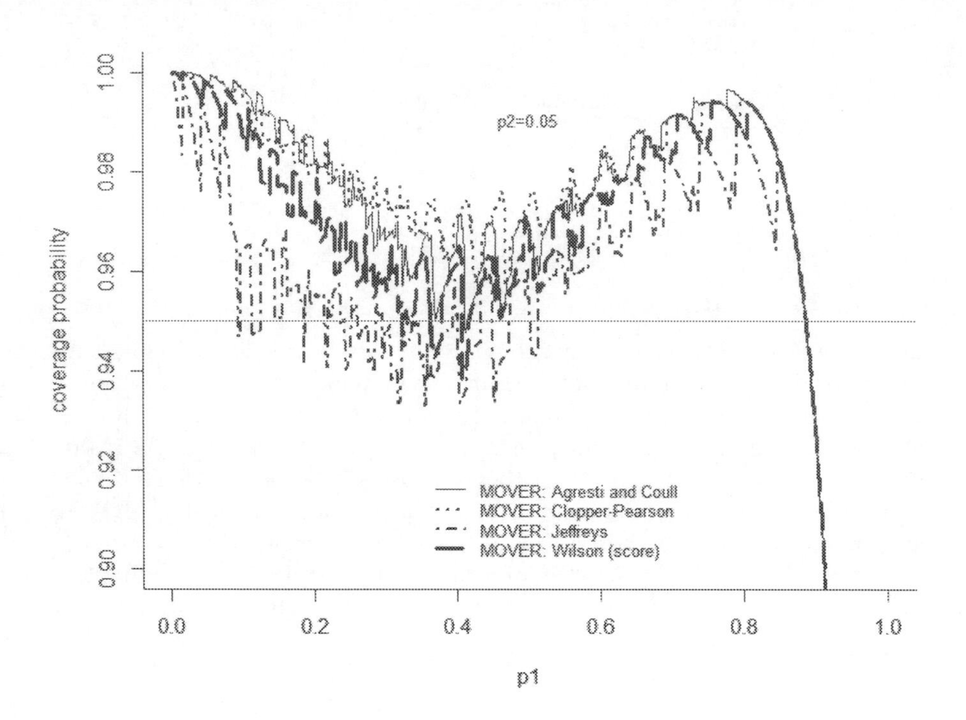

Illustrations: MOVER interval

Let us consider the data *C*D6wkRemisson introduced in section 4.2.2.2. Donner and Zou [32] recommend using Wilson interval for computing individual confidence intervals; however, in the following SAS code we use all intervals available in SAS's PROC FREQ.

```
data CD6wkRemisson;
   input Exposure Response Count;
   datalines;
   1 1 25
   1 0 182
   2 1 40
   2 0 169
   ;
data _one; set CD6wkRemisson;
/*SAS by default takes lowest value as response, coded 0 as 2
and response as 1*/
Outcome=2-Response;
run;

/*SAS's option binomial (all) computes CIs using all methods available*/
```

```
ods output BinomialCLs=_output;
proc freq data=_one;
   tables Outcome/binomial(all) alpha=0.05;
   weight Count;
  by exposure;
run;

data _out1 _out2; set _output;
   if exposure=1 then output _out2 ;
   else if exposure=2 then output _out1;
run;

/*Renaming all (l1, u1) as the CI of treatment and (l2, u2) as
the CI of placebo*/
data _out1 (rename=(Proportion=p1 LowerCL=l1 UpperCL=u1));
   set _out1; n=_n_;
data _out2(rename=(Proportion=p2 LowerCL=l2 UpperCL=u2));
   set _out2; n=_n_;run;
data _combo;
   merge _out1 _out2;
   by n;
   Delta=(p1-p2);
   Lower=delta-sqrt((p1-l1)**2+(u2-p2)**2);
   Upper=(p1-p2)+sqrt((u2-p1)**2+(p2-l2)**2);
run;
```

TABLE 4.3
SAS output: MOVER intervals

Type	Delta	Lower	Upper
Wald	0.07062	0.00122	0.12217
Wilson	0.07062	0.00055	0.11284
Agresti–Coull	0.07062	−3E05	0.11307
Jeffreys	0.07062	0.00099	0.11499
Clopper–Pearson (Exact)	0.07062	−0.0025	0.11557

Recall that the 95% Agresti–Caffo interval obtained in section 4.2.2.2 is $(0.0003, 0.1396)$. However, Agresti–Coull MOVER interval is $(-3E05, 0.1131)$ which includes 0, and is slightly shorter.

4.2.3 Exact methods

In real-world applications, especially in clinical trials, where sample sizes tend to be smaller, use of exact intervals is becoming increasingly popular. Santner et al. [80], and Fagerland, Lydersen and Laake [34] have reasoned that exact methods always guarantee nominal coverage probabilities, and the expected lengths are reasonable compared to some of the superior performing asymptotic methods. In the following, we present three commonly used exact intervals which perform well but are computationally intensive.

4.2.3.1 Chan and Zhang interval

Chan and Zhang's (**CZ**) interval [22] is obtained from the two one-sided exact tests based on the score statistic $S(\boldsymbol{x})$.

Consider the data in Table 4.1, and define $\boldsymbol{x} = (x_1, x_2)$. Using the reparametrization introduced in sub-section 4.2.2.4, the probability of observing $\boldsymbol{x}$ can be written as

$$f(\boldsymbol{x}) = f(\boldsymbol{x}|p_1, \Delta) = \prod_{j=1}^{2} \binom{n_j}{x_j} p_j^{x_j}(1-p_j)^{n_j - x_j},$$

where $p_2 = p_1 + \Delta$ (WLOG). Let $P_{p_1,\Delta}(S(\boldsymbol{x}))$ ($Q_{p_1,\Delta}(S(\boldsymbol{x}))$) be the p-value of the left (right) tailed test based on the score statistic $S(\boldsymbol{x})$ as defined below:

$$P_{p_1,\Delta}(S(\boldsymbol{x})) = \sum_{\boldsymbol{y}:S(\boldsymbol{y})\leq S(\boldsymbol{x})} f(\boldsymbol{y}|p_1,\ \Delta) \left(Q_{p_1,\Delta}(S(\boldsymbol{x})) = \sum_{\boldsymbol{y}:S(\boldsymbol{y})\geq S(\boldsymbol{x})} f(\boldsymbol{y}|p_1,\ \Delta) \right). \tag{4.28}$$

In the above equation, p_1 is a nuisance parameter. To make the p-values free of the nuisance parameter, the p-values are maximized over the permissible range of p_1. Let us define,

$$P_\Delta(S(\boldsymbol{x})) = \sup\{P_{p_1,\Delta}(S(\boldsymbol{x})) : p_1 \in I(\Delta)\}$$

and

$$Q_\Delta(S(\boldsymbol{x})) = \sup\{Q_{p_1,\Delta}(S(\boldsymbol{x})) : p_1 \in I(\Delta)\},$$

where, $I(\Delta) \equiv (\max(0, -\Delta), \min(1, 1-\Delta))$.

The $100(1-\alpha)\%$ **CZ** interval $(\underline{\Delta}^{CZ}, \overline{\Delta}^{CZ})$ is obtained as the solution of the equations

$$P_{\overline{\Delta}^{CZ}}(S(\boldsymbol{x})) = \frac{\alpha}{2} = Q_{\underline{\Delta}^{CZ}}(S(\boldsymbol{x})).$$

Figure 4.10 shows the coverage probabilities of **CZ** interval in the same setting as Figure 4.9. Note that the exact interval always guarantees the nominal coverage level (see Santner et al. [80]), and Figure 4.10 confirms the same. This interval is the most popular among all exact intervals in real-life applications (Santner et al. [80]) because the confidence limits are obtained by inverting two one-sided tests, and thus, is appropriate for testing any one-sided hypothesis.

In the following, StactXact PROCs software is used to illustrate computation of some of the exact intervals such as the **CZ** interval and Agresti and Min [2] interval. For computation of these intervals, compared to SAS's PROC FREQ, StatXact PROCs PROC BINOMIAL seems to be a bit faster. However, a word of caution, in some rare circumstances (an example is given

FIGURE 4.10
Coverage probabilities of **CZ** confidence intervals for $n_1 = 25$, $n_2 = 25$ and $p_2 = 0.05$.

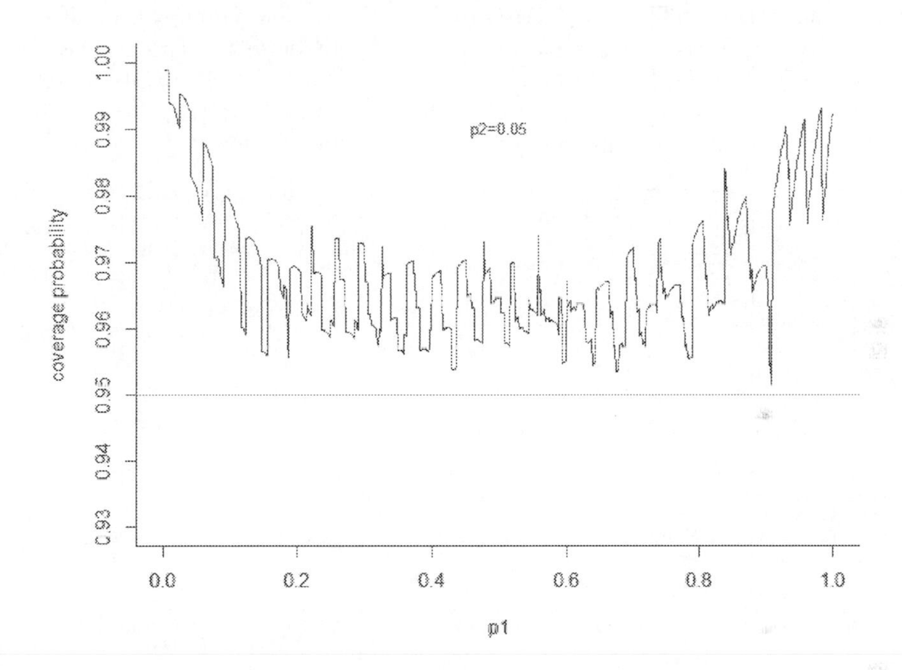

in Section 5.4), PROC BINOMIAL produces erroneous results.

Illustration: CZ interval

To illustrate **CZ** interval we use the Atlas trial data introduced in section 4.2.2.1. Since the performance of asymptotic Wald interval is questionable, we use the exact **CZ** interval. First, we run the following StatXact PROCs command using PROC BINOMIAL.

```
data DESAtlas;
   input Exposure Response Count;
   datalines;
   1 1  68
   1 0 787
   2 1  67
   2 0  889
;
PROC BINOMIAL DATA = DESAtlas gamma=0 alpha=0.9;
PD/EX ONE STD;
OU Response;
```

```
PO Exposure;
WEIGHT COUNT;
RUN;
```

In the above PROC BINOMIAL, the syntax *PD*/EX ONE STD is for proportion difference (PD) using exact (EX) method that uses ONE-sided test using standardized (STD) test statistics. Also, the option *g*amma=0 invokes no restriction in searching the parameter space. For details of this method, we refer to Berger and Boos (1994). StatXact/StatXact PROCs user manual also gives a detailed discussion of this method. Finally, the option *alpha=0.9* invokes a 90% confidence interval. The output is shown below.

```
CONFIDENCE INTERVAL ON DIFFERENCE OF TWO BINOMIAL PROPORTIONS BASED ON THE STANDARDIZED
STATISTIC AND INVERTING TWO 1-SIDED TESTS

Data file name : < DESATLAS >
Population Variable Name : Exposure
Outcome Variable Name : Response
Weight Variable Name : Count



Statistics based on the observed  2 by  2 table :

    Observed proportion for population <        1> : piHat_1          =      0.0795
    Observed proportion for population <        2> : piHat_2          =      0.0701
    Observed difference of proportions : piHat_2-piHat_1              =     -0.0094
    Stderr (pooled estimate of stdev of piHat_2-piHat_1)              =      0.0124
    Standardized difference (t): (piHat_2-piHat_1)/Stderr             =     -0.7642

Results:
--------------------------------------------------------------------------------
                             P-value                           90.00% Conf. Interval
Method           1-sided(Pr{T .LE. t})       2*1-sided              for pi_2-pi_1
--------------------------------------------------------------------------------
Asymp                 0.2224                  0.4447       (      -0.0302,       0.0109)
Exact                 0.2886                  0.5771       (      -0.0303,       0.0112)

(Note : Default value of gamma is changed to       0.0000)
```

In the above output, the lower limit of the 90% confidence interval is less than -0.03. Hence, the **CZ** interval fails to reject the null hypothesis. StatXact PROCs also offers the syntax of testing non-inferiority directly. One can use the following alternative code:

```
data DESAtlas;
   input Exposure Response Count;
   datalines;
   1 1  68
   1 0 787
   2 1  67
   2 0  889
;

/*StatXact PROCs computes CI for p2 -p1, datastep needs to be used
to change the population */
data DESAtlas1; set DESAtlas;
Exposure1=3-Exposure;
run;
/*code for non-inferiority in StatXact PROCs*/
```

```
PROC BINOMIAL DATA =DESAtlas1 gamma=0 alpha=0.9;
NONINF/EX DIFF MARGIN=0.03;
OU Response;
PO Exposure1;
WEIGHT COUNT;
RUN;
```

StatXact PROCs shows the following output:

```
UNCONDITIONAL TEST OF NON-INFERIORITY USING DIFFERENCE OF TWO BINOMIAL PROPORTIONS

Data file name : < DESATLAS1 >
Population Variable Name : Exposure1
Outcome Variable Name : Response
Weight Variable Name : Count

    H0:(pi_2-pi_1) .GE. delta_0 vs H1: (pi_2-pi_1) .LT. delta_0

Statistics based on the observed  2 by  2 table :

    Observed proportion for population <         1> : piHat_1                  =      0.0701
    Observed proportion for population <         2> : piHat_2                  =      0.0795
    Observed difference of proportions : piHat_2-piHat_1                       =      0.0094
    Maximum  margin of non-inferiority : pi_2-pi_1 = delta_0                   =      0.0300
    Stderr (restricted mle of stdev of pi_2-pi_1-delta_0 given delta_0)        =      0.0126
    Standardized test statistic (t):(piHat_2-piHat_1-delta_0)/Stderr           =     -1.6299

Results:
---------------------------------------------------------------------
                1-sided P-value                95.00% Upper Confidence
     Method      Pr{T .LE. t}                   Bound for pi_2-pi_1
---------------------------------------------------------------------
      Asymp             0.0516                        0.0302
      Exact             0.0547                        0.0303
```

The exact p-value of 0.0547 is non-significant at 5% level showing the null hypothesis $H_0 : p_1 - p_2 \leq -\Delta_0$ holds, where $\Delta_0 = 0.03$ is the chosen margin.

As stated earlier, the implementation of this method in StatXact/StatXact PROCs seems to be computationally slightly more efficient than the SAS's PROC FREQ. Using a computer with Intel Core i5 CPU 1.9GHz with 8 GB RAM, the PROC BINOMIAL of StatXact PROCs takes nearly 2.5 minutes to show the above results. On the other hand, for the above data set, which is moderately large, the following SAS code fails to produce any result.

```
data DESAtlas;
   input Exposure Response Count;
   datalines;
   1 1  68
   1 0 787
   2 1  67
   2 0  889
;
proc freq data=DESAtlas;
tables Exposure*Response/riskdiff (method=score noninf margin=0.03) alpha=0.05;
weight Count;
exact riskdiff;
run;
```

4.2.3.2 Agresti and Min interval

The Agresti and Min [2] (**AM**) interval is similar in spirit to the **CZ** interval but is based on a two-sided test for H_0: $\Delta = \Delta_0$. For each $\Delta \in (-1, +1)$, set

$$R_{p_1,\Delta}(S(\boldsymbol{x})) \quad = \sum_{\{\boldsymbol{y}:|\ S(\boldsymbol{y})| \leq |S(\boldsymbol{x})|\}} f_{p_1,\Delta}(\boldsymbol{y}).$$

The nuisance parameter p_1 is eliminated by taking the supremum of $R_{p_1,\Delta}$ over all possible values of p_1 in $I(\Delta)$ and we set

$$R_\Delta(S(\boldsymbol{x})) = \sup\{R_{p_1,\Delta}(S(\boldsymbol{x})) : p_1 \in I(\Delta)\}.$$

The $100(1-\alpha)\%$ **AM** interval $(\underline{\Delta}_{AM}, \overline{\Delta}^{AM})$ is obtained as follows. The lower limit $\underline{\Delta}_{AM}$ is the value of Δ obtained by starting at $\Delta = -1$ and increasing Δ until

$$R_{\underline{\Delta}_{AM}}(S(\boldsymbol{x})) = \alpha \ .$$

Similarly, the upper limit $\overline{\Delta}^{AM}$ is the value of Δ obtained by starting at $\Delta = 1$ and decreasing Δ until

$$R_{\overline{\Delta}^{AM}}(S(\boldsymbol{x})) = \alpha.$$

Figure 4.11 shows the coverage probabilities for varying p_1 in the same setting as Figure 4.10. Clearly, the coverage probabilities are more than the nominal level (95%) for all values of p_1. The **AM** interval guarantees the nominal coverage level (see Santner et al. [80]). Also, the length of the **AM** interval is known to be shorter than the **CZ** interval (Santner et al. [80]).

To the best of our knowledge, **AM** interval is implemented only in StatXact. Santner et al. [80] showed that in terms of coverage and expected length, this interval is the second-best. That being said, this interval cannot be used for testing one-sided hypotheses (like a one-sided hypothesis test for non-inferiority or superiority) since this interval is obtained by inverting a two-sided test. Of course, there are situations when researchers are interested in testing a two-sided hypothesis. For example, in clinical trials, serious adverse events (SAE) of a test/active compound are routinely compared with a corresponding competitor compound. In such a situation, typically, a two-sided statistical hypothesis is used. Since **AM** interval guarantees nominal coverage probability with a shorter length, it is a good candidate for computing a confidence interval of the difference of proportions.

Illustration: Agresti and Min interval

In the following, we illustrate the computation of **AM** interval using the GEMINI 3 data introduced in section 4.2.2.2. As stated earlier, in testing the remission rates at week 6 between the Vedolizumab and the placebo, Agresti and

FIGURE 4.11
Coverage probabilities of Agresti and Min confidence intervals inverting one 2-sided test for $n_1 = 25$ and $n_2 = 25$.

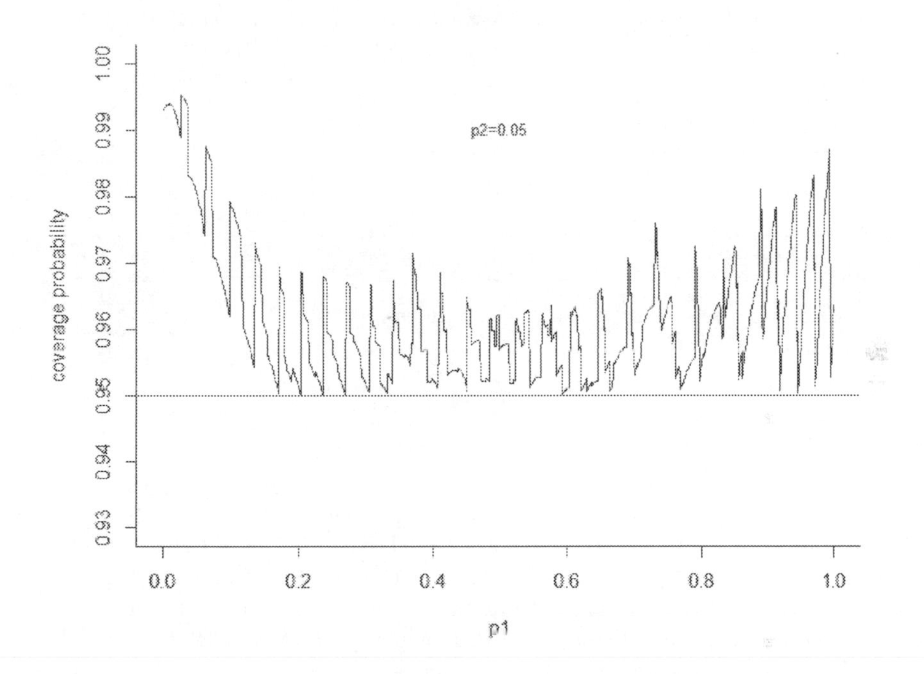

Caffo [1] method computed the 95% confidence interval as (0.0003, 0.1396), which barely excludes 0. Using the same data, let us compute the 95% **AM** confidence interval using the following code of PROC BINOMIAL.

```
data CD6wkRemisson;
input Exposure Response Count;
datalines;
1 1 25
1 0 102
2 1 40
2 0 169
;
PROC BINOMIAL DATA =CD6wkRemisson gamma=0;
PD/EX TWO STD;
OU Response;
PO Exposure;
WEIGHT COUNT;
RUN;
```

In the PROC BINOMIAL, the syntax *P*D/EX TWO STD provides the confidence interval for proportion difference (PD) using exact (EX) method that uses a two-sided (TWO) test using standardized (STD) test statistics. Also, the option *g*amma=0 invokes no Berger and Boos (1994) restriction (by default, StatXact PROCs uses gamma=1e-5). StatXact PROCs shows the following results.

```
CONFIDENCE INTERVAL ON DIFFERENCE OF TWO BINOMIAL PROPORTIONS BASED ON THE STANDARDIZED
STATISTIC AND INVERTING A 2-SIDED TEST

Data file name : < CD6WKREMISSON >
Population Variable Name : Exposure
Outcome Variable Name : Response
Weight Variable Name : Count

Statistics based on the observed  2 by  2 table :

    Observed proportion for population <        1> : piHat_1         =      0.1208
    Observed proportion for population <        2> : piHat_2         =      0.1914
    Observed difference of proportions : piHat_2-piHat_1             =      0.0706
    Stderr (pooled estimate of stdev of piHat_2-piHat_1)             =      0.0356
    Standardized difference (t): piHat_2-piHat_1)/Stderr             =      1.9833

Results:
-------------------------------------------------------------------------
                    P-value                      95.00% Conf. Interval
Method         2-sided(Pr{|T| .GE. |t|})             for pi_2-pi_1
---------------------------------------------------------------------------
Asymp               0.0473                    (      0.0008,       0.1411)
Exact               0.0481                    (      0.0003,       0.1420)
```

The above exact confidence interval shows the confidence interval $(0.0003, 0.142)$ is nearly identical to that of the Agresti and Caffo interval, the upper limit is slightly larger than the upper limit of Agresti and Caffo. The unconditional exact p-value 0.0481 shown by StatXact PROCs also confirms the marginal significance at the 5% level.

4.2.3.3 Coe and Tamhane interval

For small sample sizes, with regard to achieving the nominal coverage level and the shortest expected length, Santner et al. [80] find that Coe and Tamhane [27] (**CT**) interval performs the best. The **CT** interval is based on a computer-intensive grid search method which is described below. However, we refer the interested readers to Coe and Tamhane [27] for further details.

1. Partition the Δ-space, $[-1, +1]$, into equi-spaced intervals with the boundaries $-1 \leq \Delta_{-M} < \Delta_{-M+1} < \cdots < 0 = \Delta_0 < \Delta_1 < \cdots < \Delta_M \leq +1$, where $\Delta_{-i} = -\Delta_i$, for $1 \leq i \leq M$. The number M determines the desired accuracy of the resulting interval. Bigger the value of M, more is the accuracy. Set $i = 1$.
2. Partition the p_1-space $[\Delta_i, 1]$ into N_i intervals with boundaries $0 \leq p_{i0} < p_{i1} < \cdots < p_{iN_i}$ symmetrically about the midpoint $\frac{(1+\Delta_i)}{2}$.
3. For each $j = 0, \ldots, N_i$, construct the (nonrandomized) acceptance set A_{ij} for testing H_0: $\Delta = \Delta_i$ at α-level that contains the most

probable outcomes when $p_1 = p_{ij}$, i.e., form A_{ij} so that

$$f(\boldsymbol{x}|\, p_1, p_2) \geq f(\boldsymbol{y}|\, p_1, p_2) \quad \text{for all } \boldsymbol{x} \in A_{ij} \text{ and } \boldsymbol{y} \notin A_{ij}$$

where $p_1 = p_{ij}$, $p_2 = p_{ij} - \Delta_i$, and

$$\sum_{\boldsymbol{x} \in A_{ij}} f(\boldsymbol{x}|\, p_1, p_2) = P\{\boldsymbol{X} \in A_{ij}|\, p_1 = p_{ij}, \Delta = \Delta_i\} \geq 1 - \alpha.$$

4. Construct the (combined) acceptance region $A_i = \cup_{j=0}^{N_i} A_{ij}$ of H$_0$: $\Delta = \Delta_i$.
5. If there is a hole in A_i, in the sense $(x_1 - 1, x_2) \in A_i$ and $(x_1 + 1, x_2) \in A_i$ but $(x_1, x_2) \notin A_i$, Add $\boldsymbol{x}$ to A_i to eliminate the hole. In other words, add points to eliminate "holes" in either the x_1-direction or in the x_2-direction.
6. Let $\widehat{\Delta}(\boldsymbol{x}) = x_1/n_1 - x_2/n_2$. Eliminate any $\boldsymbol{x}^*$ from A_i for which $\widehat{\Delta}(\boldsymbol{x}^*) \leq \min_{x \,\in\, A_{i-1}} \widehat{\Delta}(\boldsymbol{x})$ and $\boldsymbol{x}^* \notin A_{i-1}$.
7. For each $0 \leq i \leq M$, let

$$P\{A_i|\, \Delta_i\} \equiv \inf_{p_1 \in I(\Delta_i)} P\{\boldsymbol{X} \in A_i|\, p_1, \Delta_i\}.$$

then $P\{A_i|\, \Delta_i\} > 1 - \alpha$ by construction. Also let

$$\mathcal{D} = \{\boldsymbol{x} \in A_i| \text{ and } P\{A_i - \{\boldsymbol{x}\}|\, \Delta_i\} \geq 1 - \alpha\ \}.$$

(When $n_1 = n_2$ then $\mathcal{D}$ is modified to be the set of pairs of points

$$\begin{aligned} \mathcal{D} &= \{(\boldsymbol{x}, \boldsymbol{n} - \pi\boldsymbol{x})|\{\boldsymbol{x}, \boldsymbol{n} - \pi\boldsymbol{x}\} \\ &\in A_i \ \&\ P\left(A_i - \{\boldsymbol{x}, n - \pi\boldsymbol{x}\}, |\Delta_i\right) \geq 1 - \alpha\}) \end{aligned}$$

Here $\pi(x_1, x_2) = (x_2, x_1)$ is the permutaion of $\boldsymbol{x}$. The separate definition for $n_1 = n_2$ ensures that the final intervals have certain invariance properties. The set $\mathcal{D}$ can be thought of as "candidate points" for elimination from the acceptance set A_i. Eliminate the (myopically optimal) point $x^* \in \mathcal{D}$ from A_i where $P\{A_i - \{x^*\}|\, \Delta_i)\} = \max_{x \in \mathcal{D}} P\{A_i - \{x\}|\, \Delta_i\}$. (If $n_1 = n_2$ then pairs of points $(\boldsymbol{x}, \boldsymbol{n} - \pi\boldsymbol{x})$ are eliminated from A_i in a similar fashion.)
8. Construct acceptance sets $A_{-i} = \{\boldsymbol{n} - \boldsymbol{x}|\ \boldsymbol{x} \in A_i\}$ corresponding to H$_0$: $\Delta = \Delta_{-i}$ for $i = 1, \ldots, M$.
9. Invert the acceptance sets $\{A_i\}$ to form the the confidence interval $\left(\underline{\Delta}^{CT}, \overline{\Delta}^{CT}\right)$ at $\mathbf{x}$ as follows

$$\begin{aligned} \underline{\Delta}^{CT}(\boldsymbol{x}) &= \min_{-M \leq i \leq M} \{\Delta_i : x \in A_i\} \text{ and } \overline{\Delta}^{CT}(\boldsymbol{x}) \\ &= \max_{-M \leq i \leq M} \{\Delta_i : x \in A_i\}. \end{aligned}$$

Due to the complexity of the algorithm, **C**T interval seems difficult to implement, and to the best of our knowledge, it has not been available in any commercially available software. A SAS macro implementation of this method using SAS 6 may be available from the authors, and a C++ implementation of this method (not tested thoroughly) is available from the website for the book (`http://math.bu.edu/people/ag/`).

Figure 4.12 shows the coverage probabilities for varying p_1 values under the same setting as Figure 4.11. As shown in the figure, this exact method shows excellent coverage probabilities. Santner et al. [80] showed that in terms of coverage and expected length properties, this is the best two-sided exact confidence interval.

FIGURE 4.12
Coverage probabilities of Coe and Tamhane confidence intervals for $n_1 = 25$ and $n_2 = 25$.

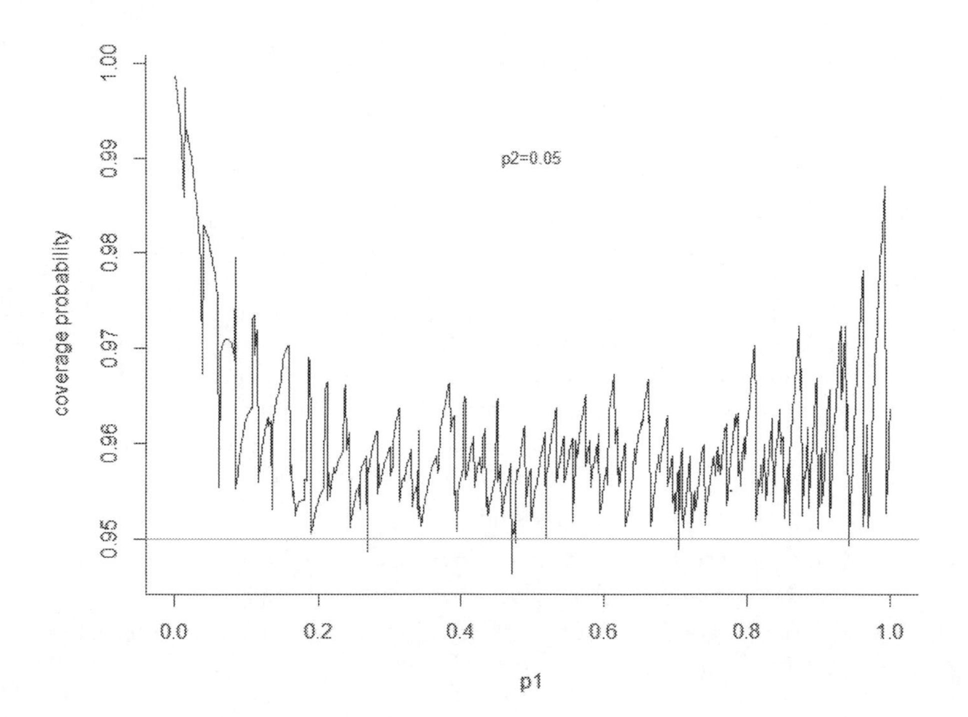

Santner and Yamagami [82] also propose a similar approach to construct confidence intervals based on the acceptance region. Computer code implementing this method is available from the authors. Santner et al. [80] show that even though Santner and Yamagami [82] exact method preserves the nominal coverage probability, the expected lengths are much higher than the other exact methods, hence is not recommended for use in practice. Santner and Snell [81] also propose a method for computing exact confidence intervals based on unstandardized test statistics $\widehat{p}_1 - \widehat{p}_2$. To our knowledge, this method has been implemented in SAS's PROC FREQ and StatXact PROCs software. However, the confidence interval using unstandardized test statistics tends to be very conservative (wide) and is not recommended in practice.

4.3 Bayesian intervals

This section briefly discusses the Bayesian computation of the credible intervals of the difference of two independent binomial proportions. It's worth mentioning that we adopt the Bayesian method as an alternative technique to generate an interval estimate of the parameter of interest. Like other competing intervals, we use the coverage probability and the expected length to measure its performance.

Let's define $\boldsymbol{\theta} = (p_1, p_2)$, and $h(\boldsymbol{\theta})$ be a real-valued function of $\boldsymbol{\theta}$, the parameter of interest. In order to find a credible interval of $h(\boldsymbol{\theta})$, the first step is to compute the posterior density of $\boldsymbol{\theta}$, say, $g(\boldsymbol{\theta}|\boldsymbol{x})$ where, $\boldsymbol{x} = (x_1, x_2)$ (cf. Section 1.3). The posterior density of $h(\boldsymbol{\theta})$ can then be easily obtained from $g(\boldsymbol{\theta}|\boldsymbol{x})$. A $(1-\alpha)$ credible interval $C(x)$ for the parameter $h(\boldsymbol{\theta})$ is then given by

$$P\left(h(\boldsymbol{\theta}) \in C(x)|X = x\right) = 1 - \alpha. \tag{4.29}$$

Let h_α be the $(1-\alpha)$ percentile of the posterior distribution of $h(\boldsymbol{\theta})$, i.e.,

$$P\left(h(\boldsymbol{\theta}) \leq h_\alpha|X = x\right) = 1 - \alpha, \tag{4.30}$$

then an equal-tailed $(1-\alpha)$ credible interval of $h(\boldsymbol{\theta})$ is given by $(h_{\alpha/2}, h_{1-\alpha/2})$.

Let's consider $h(\boldsymbol{\theta}) = p_1 - p_2$, the difference of two binomial proportions, as the parameter of interest. The posterior density of $\boldsymbol{\theta}$ is given by

$$g(\boldsymbol{\theta}|\boldsymbol{x}) \propto f(\boldsymbol{x}|\boldsymbol{\theta})\pi(\boldsymbol{\theta}) \tag{4.31}$$

where, $f(\boldsymbol{x}|\boldsymbol{\theta})$ is the likelihood of the data, and $\pi(\boldsymbol{\theta})$ is a prior distribution of $\boldsymbol{\theta}$.

We present here the computation of a credible interval of $h(\boldsymbol{\theta}) = p_1 - p_2$ using the conjugate prior distribution $Beta(\alpha, \beta)$. Note that

$$f(\boldsymbol{x}|\boldsymbol{\theta}) = \prod_{j=1}^{2} \binom{n_j}{x_j} p_j^{x_j} (1-p_j)^{n_j - x_j}, \tag{4.32}$$

and

$$\pi(\boldsymbol{\theta}) = \prod_{j=1}^{2} \frac{\Gamma(\alpha_j + \beta_j)}{\Gamma(\alpha_j)\Gamma(\beta_j)} p_j^{\alpha_j - 1} (1-p_j)^{\beta_j - 1}. \tag{4.33}$$

Thus, the posterior density $g(\boldsymbol{\theta}|\boldsymbol{x})$ becomes,

$$\begin{aligned} g(\boldsymbol{\theta}|\boldsymbol{x}) &\propto \prod_{j=1}^{2} \left[\binom{n_j}{x_j} p_j^{x_j} (1-p_j)^{n_j - x_j} \times Beta(\alpha_j, \beta_j) \right] \\ &= \prod_{j=1}^{2} \frac{\Gamma(\alpha_j + \beta_j)}{\Gamma(\alpha_j)\Gamma(\beta_j)} \binom{n_j}{x_j} p_j^{(x_j + \alpha_j - 1)} (1-p_j)^{(n_j - x_j + \beta_j - 1)} \\ &= \prod_{j=1}^{2} \left[Beta(x_j + \alpha_j, n_j - x_j + \beta_j) \right]. \end{aligned} \tag{4.34}$$

Thus, the joint posterior distribution is the product of two marginal posterior distributions $Beta(x_j + \alpha_j, n_j - x_j + \beta_j)$, $j = 1, 2$. In Bayesian inference, often Markov Chain Monte Carlo (MCMC) methods are used to generate samples from a posterior distribution. Using Gibbs sampling, a form of MCMC method, one can easily generate a sample $p_j^1, p_j^2, .., p_j^N$ from the marginal posterior distribution $Beta(x_j + \alpha_j, n_j - x_j + \beta_j)$ of p_j, $j = 1, 2$. In turn, $p_1^1 - p_2^1, p_1^2 - p_2^2, .., p_1^N - p_2^N$ represent a random sample from the posterior distribution of $h(\boldsymbol{\theta}) = p_1 - p_2$. The level $(1 - \alpha)$ credibility interval of $h(\boldsymbol{\theta})$ is, thus, computed from the sample percentiles, viz., the percentiles of $p_1^1 - p_2^1, p_1^2 - p_2^2, .., p_1^N - p_2^N$. Clearly, the Beta prior depends on the choices of α_j and β_j (cf. Figure 1.3), which, in turn, influences the posterior density. In particular, the prior corresponding to $\alpha_j = \beta_j = 1$ is the uniform density, while for $\alpha_j = \beta_j = 0.5$, we get what is known as non-informative Jeffreys prior.

Once the random samples are generated from the marginal posteriors of p_1 and p_2, the computation of the credible interval of any function $h(\boldsymbol{\theta})$ of p_1 and p_2 is straightforward. For example, if $h(\boldsymbol{\theta}) = p_1 - p_2$ is the parameter of primary interest, then the following PROC MCMC code can be used to compute the credible interval. Let's consider the TAXUS ATLAS trial [92] data introduced in section 4.2.2.1. First, we enter the data using the data step given below and then run PROC MCMC using the following syntax:

```
data one;
input trt y total;
datalines;
1 67 956
```

```
2 68 855
;

ods graphics on;
proc mcmc data=one seed=12346 nmc=100000 nbi=500 outpost=out1 nthin=10 monitor=(delta);
   array p[2] p1 p2;
   parms p1 p2;
   prior p1 p2 ~beta(0.5,0.5);
   model y ~ binomial(total, p[trt]);
   beginnodata;
   delta=p1-p2;
   endnodata;
run;
ods graphics off;
```

In the above PROC MCMC code, $n$mc=100000 specifies the number of samples to be generated, $n$bi=500 option invokes the burn-in to be used, and $n$thin=10 option specifies the thinning to be applied. In the PROC MCMC code, the *array p[2] p1 p2* defines the two parameters p_1 and p_2. The option $p$arms identify the parameters for which marginals to be computed. The option *prior p1 p2 beta(0.5,0.5)* specifies the priors $Beta(0.5, 0.5)$ corresponding to the parameters. The option *model y binomial(total, p[trt])* invokes likelihood as $Binomial(N_i, p_i)$ with N=total and $p_i = p[trt]$. Finally, the command *delta=p1-p2* within *beginnodata* and *endnodata* computes the difference *delta* of the two marginals, and the option *monitor=(delta)* displays the output related to *delta*.

Figure 4.13 exhibits SAS's MCMC output aids for convergence checks of Markov chains for the marginal distributions of p_1 and p_2, and also plots of its' posterior distributions. The top panel shows the trace plot which appears to be centred around a constant value with small fluctuation. It indicates the convergence of the chain to the right stationary distribution. The plots in the Autocorrelation panel show the presence of a small degree of autocorrelation at the beginning of the sampling, and then it diminishes rapidly. We may thus conclude that the mixing is quite good. Finally, the posterior density plot of the marginals are shown in the panel named 'Posterior Density'.

Figure 4.14 displays the diagnostics for 'delta', which is derived from the marginals given in Figure 4.13. Since the trace plot shows good mixing and other diagnostic plots show no apparent issues, one can assume that all chains have converged. SAS's PROC MCMC shows the following as the posterior summary.

Posterior Summaries and Intervals

Parameter	N	Mean	Standard Deviation	95% HPD Interval	
p1	10000	0.0706	0.00828	0.0544	0.0865
p2	10000	0.0801	0.00925	0.0620	0.0979
delta	10000	-0.00954	0.0124	-0.0336	0.0152

FIGURE 4.13
Bayesian diagnostics of the marginal distributions from PROC MCMC.

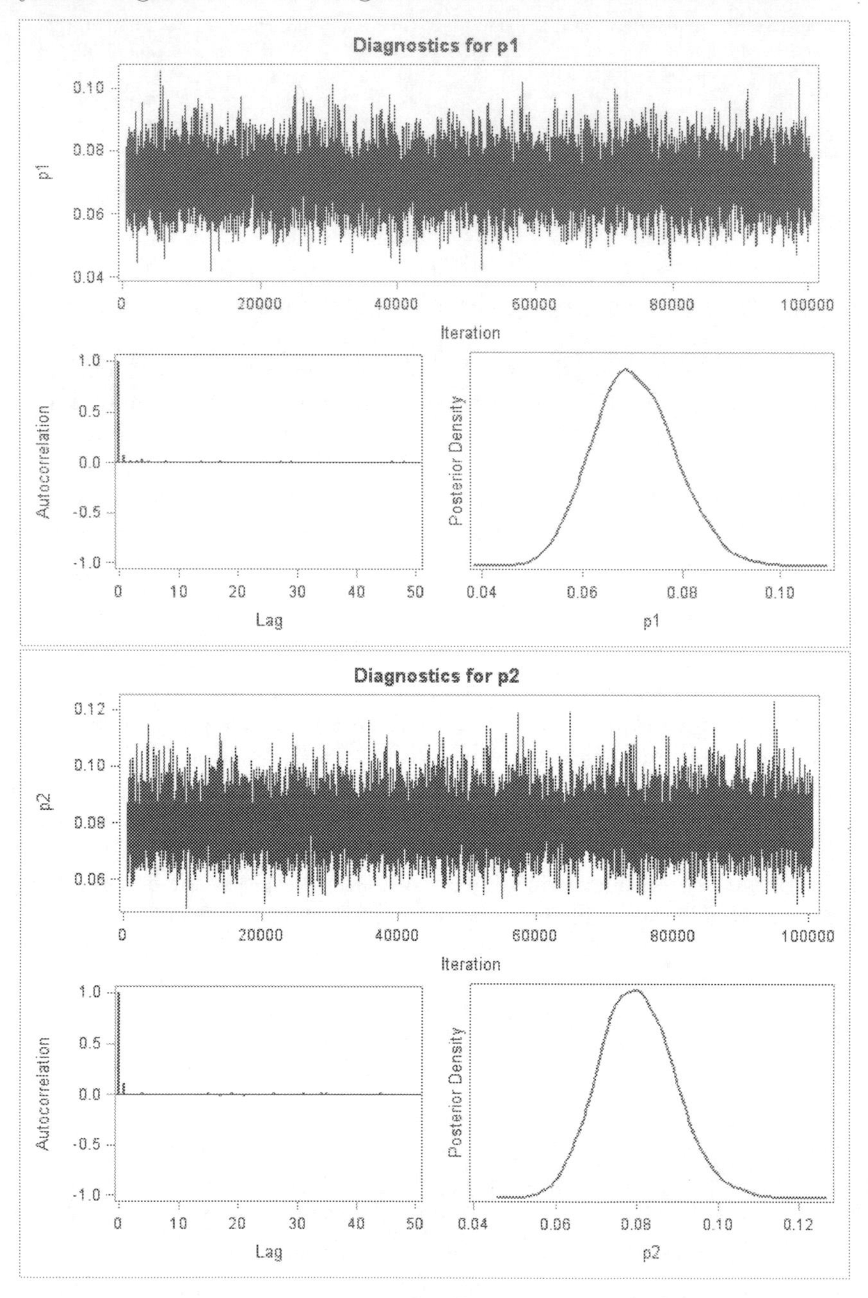

FIGURE 4.14
Bayesian diagnostics for $p_1 - p_2$ from PROC MCMC.

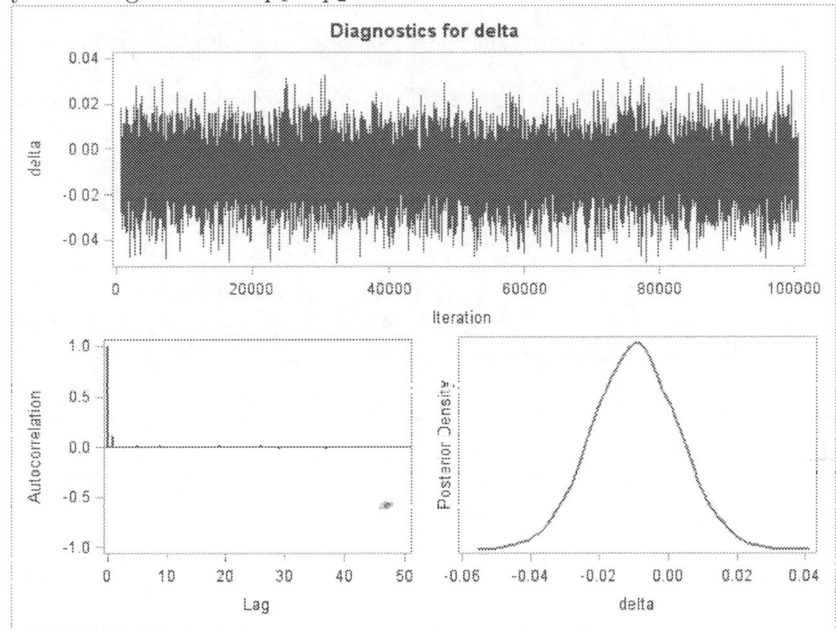

In the above 95% HPD interval (highest posterior density interval) for delta, which is $p_1 - p_2$, is found to be $(-0.0336, 0.0152)$. Note that the HPD interval is narrower than the usual equal tail credible interval.

4.3.1 Discussion and recommendation

In this chapter, we discuss the formulations of the commonly used asymptotic and exact confidence intervals for the difference of two independent binomial proportions. We also provide the SAS/R codes for its' implementation. Also, we illustrate its' real-life applications using some live data sets. These examples clearly demonstrate that, use of different intervals may lead to different inferences, especially when the sample sizes are small. Thus, in order to make a judicious choice of what interval to use in a given situation, a practitioner needs to weigh the pros and cons of using different intervals in terms of its' achieved coverage and expected length.

Evidently, there is no one size fits all solution. Newcombe ([60]) compares the performances of eleven asymptotic two-sided confidence intervals. Among these intervals, the following intervals are recommended based on its' attained coverage, expected lengths and locations.

- Miettinen and Nurminen

- True Profile
- Wilson's score – stated as Newcombe
- Wilson's score with continuity correction – Newcombe (continuity corrected)

Thus, Miettinen and Nurminen (**MN**) interval is considered to be the most preferred interval that performs well. Santner et al. [80] compare this interval with four other exact intervals – viz **CZ**, **AM**, **CT**, and **SY** (Santner and Yamagami [82]). They find that, for small sample sizes, the **MN** interval fails to achieve the nominal coverage probability in more than 50% cases. More recently, Newcombe and Nurminen [61] point out that Santner et al. [80] fail to include the correction factor $\frac{N}{N-1}$ in the computation of variance of the standardized test statistic, and thus, the **MN** intervals are narrower and fail to achieve the nominal coverage probability.

Pradhan and Banerjee [68] propose a weighted profile likelihood-based (**PF**) confidence interval for the difference of two binomial proportions. They find that (**PF**) interval performs better than the *True Profile* and the *Wilson's score* intervals that Newcombe ([60]) recommend.

Santner et al. [80] find that in terms of attaining nominal coverage probability and achieving shorter expected length, among the exact intervals, the **CT** interval performs the best followed by the **AM**, and the **CZ** intervals, respectively. Since the intervals **CT** and **AM** are constructed by inverting two-sided tests they are not suitable for testing a one-sided statistical hypothesis. In the above, the performances of the exact intervals are discussed and compared. In the following, we compare the performances of the asymptotic intervals in terms of its' attained coverage and expected length.

For a given (p_1, p_2), the coverage $\zeta(p_1, p_2)$ and the expected length $\xi(p_1, p_2)$ of an interval $[\underline{\Delta}(\mathbf{x}), \overline{\Delta}(\mathbf{x})]$ are computed as follows:

$$\zeta(p_1, p_2) = \sum_{\boldsymbol{x} \in \Gamma} \prod_{j=1}^{2} \binom{n_j}{x_j} p_j^{x_j} (1 - p_j)^{n_j - x_j},$$

where, $\Gamma = \{\boldsymbol{x} = (x_1, x_2) : \underline{\Delta}(\boldsymbol{x}) \leq p_2 - p_1 \leq \overline{\Delta}(\boldsymbol{x})\}$, and

$$\xi(p_1, p_2) = \sum_{\boldsymbol{x}} \prod_{j=1}^{2} \binom{n_j}{x_j} p_j^{x_j} (1 - p_j)^{n_j - x_j} \widehat{\lambda}(\boldsymbol{x}),$$

where, $\widehat{\lambda}(\boldsymbol{x}) = \overline{\Delta}(\boldsymbol{x}) - \underline{\Delta}(\boldsymbol{x})$ is the length of the computed confidence interval.

We now present the results of a simulation study mimicking a phase 2 clinical trial with $(n_1, n_2) = (65, 65)$. We partition the unit square $\{(p_1, p_2)|0 \leq p_1, p_2 \leq 1\}$ into $100 \times 100 = 100^2$ grids of equal size. Given that the grid size is sufficiently small, we may, thus, assume that each grid represents the probability (p_1, p_2), say, the centre of the grid. For each of these points (p_1, p_2) we compute $\zeta(p_1, p_2)$ and $\xi(p_1, p_2)$. Finally, the distribution of the coverage and expected lengths corresponding to these 100^2 grid points are plotted using BliP plots (Lee and Tu [53]). Figures 4.15 and 4.16 present the coverage probability distributions and the expected length distributions, respectively, of different intervals. The vertical bars inside the double triangular-shaped object corresponding to each interval show the deciles of the distributions of the achieved coverage and the expected length. A long vertical line in Figure 4.15 indicates the 95% level.

Figure 4.15 shows that, among all the intervals, **NW** (Wilson's interval due to Newcombe) performs the worst; more than 40% of the points have less than 95% nominal coverage probabilities, and there are a few points whose coverage probabilities are as low as 0.88. The continuity corrected Newcombe method, labeled as **NWC**, on the other hand, is very conservative, all the decile bars are well above the nominal 95% level, leading to much wider confidence intervals. In an earlier study, Santner et al. [80] find that 50% of the attained coverage values of **MN** interval are below the nominal 95% level. We compute all the confidence intervals using SAS's PROC FREQ that multiplies the correction factor $N/(N-1)$ to the variance in calculating the test statistics. Our simulation study results confirm the findings of Santner et al. [80]. The decile bars of **FM** are highly scattered, and most decile bars are on the right of the 95% vertical line indicating that the interval is wider. The **AC** and **PF**, on the other hand, seem to perform better. Both these methods show nearly 30% under-coverage, but most of the decile bars are close to the 95% line, indicating that these intervals are narrower. The expected length distributions of these intervals shown in Figure 4.16 support the above findings.

Hence, in terms of the coverage and expected length, the methods **AC** and **PF** performed well, and we recommend these methods to use in practice.

As stated at the end of section 4.2.2.7, even though StatXact/StatXact PROCs's PROC BINOMIAL is computationally a bit faster than SAS's PROC FREQ, on some rare occasions, it may produce erroneous results, which primarily due to the default setting of the inbuilt grid search algorithm. For example, consider the following data from an Inflammatory Bowel Disease (IBD) trial treating Ulcerative colitis (UC). The endpoint of interest was the clinical response at the end of the induction phase. The observed rates for the placebo and the treatment low dose were 21.9% (16/73) and 37.4% (26/70). The data are given in Table 4.4.

FIGURE 4.15
Distribution of the coverage probabilities of 95% confidence intervals for $n_1 = 65$ and $n_2 = 65$. AC = Agresti and Caffo FM=Farrington and Manning MN = Miettinen and Nurminen NW = Newcombe's method due to Wilson NWC = Continuity corrected NW method PF = Weighted profile likelihood.

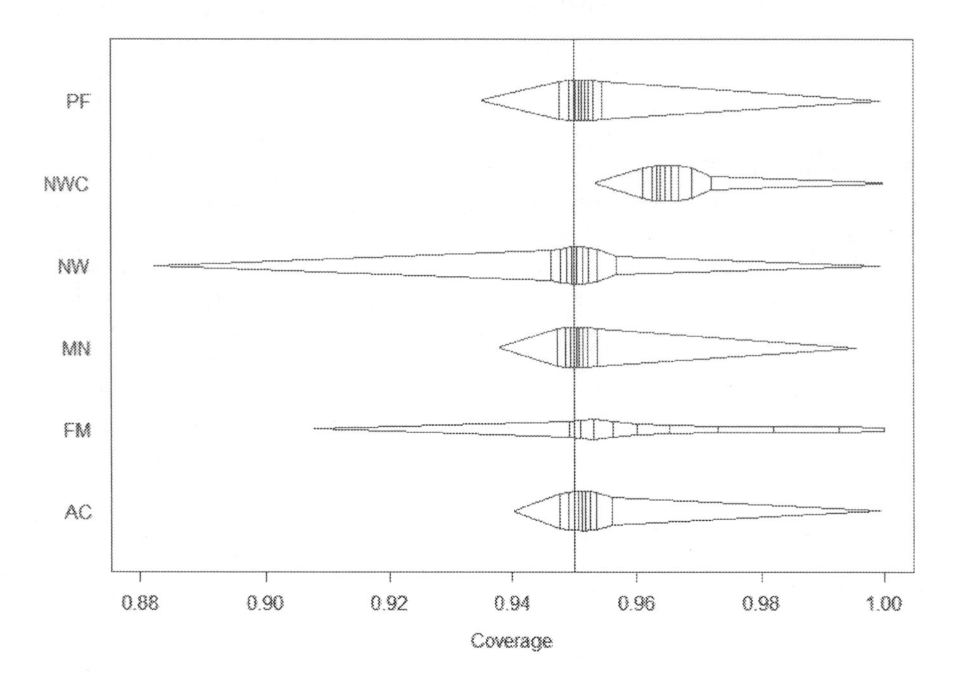

The following PROC BINOMIAL code computing the confidence interval for the difference of proportions using the Chan and Zhang method shows a statistically significant p-value 0.0485, however, the corresponding 95% confidence interval $(-0.002, 0.301)$, which includes 0, gives statistically non-significant inference.

```
data UCIBD;
input Treat Response count @@;
datalines;
1 1 26 1 2 44
0 1 16 0 2 57
;

PROC BINOMIAL data=UCIBD  Gamma=0 ;
   PD / EX ONE STD;
   PO Treat;
   outcome Response;
   weight count ;
```

FIGURE 4.16
Distribution of the expected lengths of 95% confidence intervals for $n_1 = 65$ and $n_2 = 65$. AC = Agresti and Caffo FM = Farrington and Manning MN = Miettinen and Nurminen NW = Newcombe's method due to Wilson NWC = Continuity corrected NW method PF = Weighted profile likelihood.

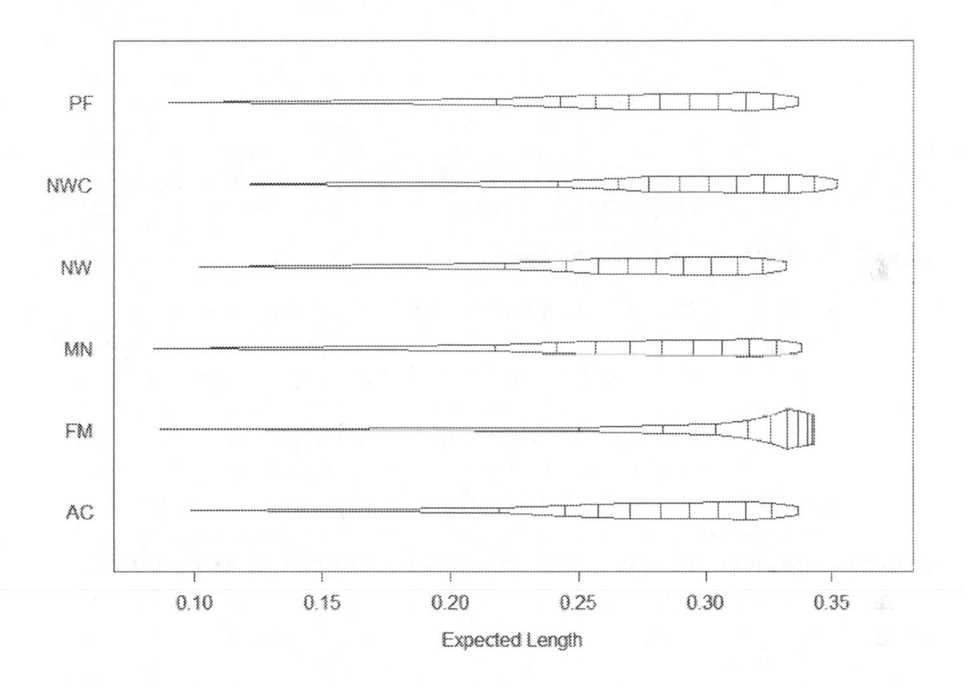


```
run ;

CONFIDENCE INTERVAL ON DIFFERENCE OF TWO BINOMIAL PROPORTIONS BASED ON THE STANDARDIZED
STATISTIC AND INVERTING TWO one-sided TESTS

Data file name : < UCIBD >
Population Variable Name : Treat
Outcome Variable Name : Response
Weight Variable Name : count


Statistics based on the observed  2 by  2 table :

    Observed proportion for population <        1> : piHat_1           =       0.2192
    Observed proportion for population <        2> : piHat_2           =       0.3714
    Observed difference of proportions : piHat_2-piHat_1               =       0.1523
    Stderr (pooled estimate of stdev of piHat_2-piHat_1)               =       0.0762
    Standardized difference (t): (piHat_2-piHat_1)/Stderr              =       1.9983
```

TABLE 4.4
Ulcerative colitis (UC) clinical trial data

	Placebo	Treatment
Yes	16	26
No	57	44
Total	73	70

```
Results:
---------------------------------------------------------------------------------
                              P-value                         95.00% Conf. Interval
Method          1-sided(Pr{T .GE. t})      2*1-sided           for pi_2-pi_1
---------------------------------------------------------------------------------
Asymp                0.0228                  0.0457      (      0.0024,       0.2977)
Exact                0.0243                  0.0485      (     -0.0021,       0.3008)

(Note: Default value of gamma is changed to       0.0000)
```

The above 95% confidence interval due to Chan and Zhang is clearly inconsistent with the corresponding p-value. There are several ways; one can check the validity of the above confidence interval. For example, using the same software, one could run the confidence interval for the ratio of proportions and check the consistency between the p-value and the corresponding confidence interval. The more advanced approach could be using the lower limit as the margin and then using the PROC BINOMIAL's non-inferiority test, and check if the one-sided p-value is 0.025. Any of these methods will confirm the computational abnormality. Therefore, users need to use a judicious approach before using such results.

5

Two Independent Binomials: Ratio of Proportions

5.1 Introduction

In this chapter, we discuss the confidence intervals of ratio of two binomial proportions, also referred to as the Relative Risk or Risk Ratio. These are often used in real-world applications, such as oncology or vaccine trials, where the non-inferiority of an experimental drug/vaccine is compared to a control drug/vaccine by a certain margin in terms of the ratio of proportions. First, we state the three hypotheses that are commonly tested. We then discuss the methodologies for computing the confidence intervals of ratio of proportions that are available in the literature with illustrative examples from real-world applications. Finally, we conclude the chapter with discussion and recommendation.

Consider two independent binomial variables X_1 and X_2 with parameters (n_1, p_1) and (n_2, p_2), respectively. A typical observed data can be represented as Table 4.1. Let $\phi = \frac{p_1}{p_2}$ with $0 \leq \phi < \infty$, the ratio of the proportions, be the parameter of interest. Therefore, an estimate of ϕ can be computed as $\widehat{\phi} = \frac{\widehat{p}_1}{\widehat{p}_2}$ where $\widehat{p}_i = \frac{x_i}{n_i}, i = 1, 2$. Let $(\underline{\phi}, \overline{\phi})$ denote the level $(1 - \alpha)$ confidence interval of ϕ. Also, if the population labels are interchanged, then given $(\underline{\phi}, \overline{\phi})$, the confidence interval of p_1/p_2, the confidence interval of p_2/p_1 is given by

$$(1/\overline{\phi}, 1/\underline{\phi}) \tag{5.1}$$

5.2 Hypotheses about the ratio of proportions

The three hypotheses are:

$$H_0 : \phi = \phi_0 \quad \text{versus} \quad H_1 : \phi \neq \phi_0 (\text{Two-sided alternatives}) \tag{5.2}$$

DOI: 10.1201/9781315169859-5

or

$$H_0 : \phi \leq \phi_0 \quad \text{versus} \quad H_1 : \phi > \phi_0 (\text{One-sided alternatives}) \tag{5.3}$$

or

$$H_0 : \phi \geq \phi_0 \quad \text{versus} \quad H_1 : \phi < \phi_0 (\text{One-sided alternatives}) \tag{5.4}$$

As discussed before (cf. Chapters 1–4), in clinical trial applications, for testing the **non-inferiority** or the **superiority** of a drug compared to a standard drug, one-sided alternatives are considered. In testing for **non-inferiority** (**Superiority**), the objective is to show that the test drug is not inferior (superior) to a standard drug by a certain margin, say ϕ_0. Suppose p_1 and p_2 represent the proportions for the test drug and the standard drug, respectively. Then, testing for the non-inferiority (superiority) can be framed as,

$$H_0 : p_1/p_2 \leq \phi_0 \quad \text{versus} \quad H_1 : p_1/p_2 > \phi_0 \tag{5.5}$$

or

$$H_0 : \phi \leq \phi_0 \quad \text{versus} \quad H_1 : \phi > \phi_0, \tag{5.6}$$

where H_1 represents the hypothesis of interest, i.e., the hypothesis of non-inferiority (superiority). Clearly, for testing the non-inferiority (superiority), the margin ϕ_0 should be less (more) than 1 (see Chow, Shao and Wang [25]).

Often, other than testing for non-inferiority or superiority of a drug, testing equivalence of two drugs is also very common. It is used to demonstrate the similarities or lack of dissimilarities between the two drugs given the pre-specified margins, say, ϕ_l and ϕ_u ($\phi_l < \phi_u$). Testing for equivalence problem can then be framed as testing

$$H_0 : \phi \leq \phi_l \quad \text{or} \quad \phi \geq \phi_u \tag{5.7}$$

versus the alternatives

$$H_1 : \phi_l < \phi < \phi_u, \tag{5.8}$$

where ϕ_l and ϕ_u are the lower and the upper margins. The rejection of the null hypothesis leads to the acceptance of the hypothesis of equivalence, that is the two ratios (relative risks) are equivalent. Later in this chapter, we illustrate the testing of these hypotheses using real-world data.

5.2.1 Asymptotic methods

Here, we discuss some commonly used asymptotic confidence intervals of $\phi = \frac{p_1}{p_2}$. Notice that the construction of these intervals relies on the normal approximation to the distribution of $\widehat{\phi}$.

5.2.1.1 Katz et al. (KZ) interval

Using the delta method, one can approximate the distribution of $log(\widehat{\phi})$ by a normal distribution with mean ϕ and variance $x_1^{-1} + x_2^{-1} - n_1^{-1} - n_1^{-1}$. Based on this approximation, Katz et al. [49] propose the following $(1-\alpha)$-confidence interval for $log(\phi)$:

$$log(\widehat{\phi}) \pm z_{\alpha/2}\sqrt{\frac{1}{x_1} + \frac{1}{x_2} - \frac{1}{n_1} - \frac{1}{n_2}}. \tag{5.9}$$

Thus, one can compute the confidence interval of the ϕ by exponentiating the limits. However, it is important to note that for $x_1 = 0$ or $x_2 = 0$, the standard error is undefined, and for $n_1 = x_1$ and $n_2 = x_2$, the standard error is 0. To address these issues, one may simply add 0.5 successes to each group (Walter [96]). Therefore, the modified estimate of ϕ is given by,

$$\widetilde{\phi} = \frac{(x_1 + 0.5)/(n_1 + 0.5)}{(x_2 + 0.5)/(n_2 + 0.5)}. \tag{5.10}$$

Thus, the continuity corrected interval of $log(\phi)$ is:

$$log(\widetilde{\phi}) \pm z_{\alpha/2}\sqrt{\frac{1}{x_1 + 0.5} + \frac{1}{x_2 + 0.5} - \frac{1}{n_1 + 0.5} - \frac{1}{n_2 + 0.5}}. \tag{5.11}$$

Finally, inverting back to the original scale, one can compute the confidence limits of ϕ. Note that this approach works even when $x_1 = x_2 = 0$. However, for $n_1 = x_1$ and $n_2 = x_2$, this method still produces a non-sensical confidence interval with both its limits equal to 1. Also, if one of the populations has zero successes, the confidence interval may not contain the point estimate $\widehat{\phi}$.

In real-world applications, such as in clinical trial, usually, the values of ϕ bigger than 1 is of interest. Mimicking a small sample setup, frequently used in clinical trials, we consider $n_1 = n_2 = 20$ and $\phi = 1.5$ (representing a situation where the success rate of the test drug is 50% more than the standard drug). The simulated coverage probability distribution of the continuity corrected interval proposed by Katz et al. is plotted for different values of p_1 in Figure 5.1.

From Figure 5.1, it is evident that the interval fails to attain the nominal coverage probability for values of p_1 greater than 0.78, thus, leading to under-coverage. On the other hand, for all values of p_1 less than 0.78, the attained coverage probabilities are more than 0.95, and thus leading to over-coverage.

FIGURE 5.1
Coverage probabilities of continuity corrected Katz et al. confidence intervals.

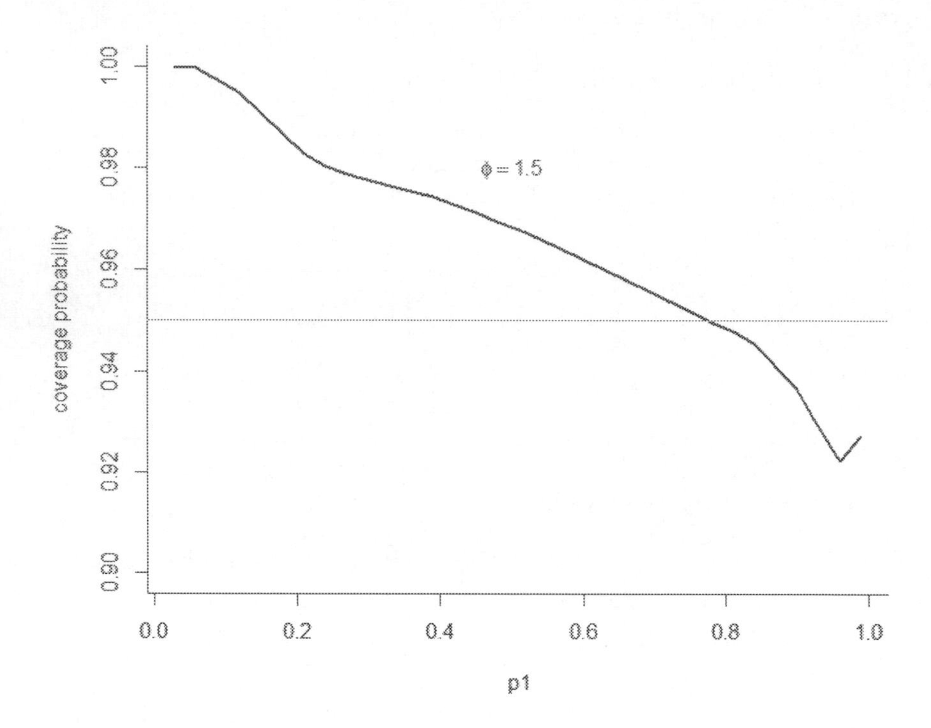

Of course, the perfect situation would be the attained coverage probabilities would coincide with the 0.95 line. Note that under-coverage leads to a narrower/liberal intervals while the over-coverage to the wider/conservative intervals. The distribution of simulated coverage probabilities in $p_1 - p_2$-space is given in the discussion section.

Illustration

To illustrate the above methods, let us consider the data from a randomized control trial (RCT) discussed in Perondi et al. [67] (also see, Fagerland, Lydersen and Laake [34]). The trial considers the following question (Perondi et al. [67]), "children who remain in cardiac arrest after cardiopulmonary resuscitation are administered with an initial standard dose of epinephrine. If resuscitation is unsuccessful, should the next dose be the same dose or a higher dose?" A sample of 34 patients was randomly administered the standard-dose, and the other 34 patients were assigned to the higher dose. The primary outcome of interest was the 24-hour survival rate after the cardiac arrest. The results of this study are given in Table 5.1.

TABLE 5.1
Pediatric cardiac arrest data with epinephrine treatment

Survival	Standard Dose	High Dose
Yes	7	1
No	27	33
Total	34	34

The SAS code (SAS's PROC FREQ) for computing interval of Katz et al. is given below.

```
ods listing;
data predcardiac1;
input Dose Response Count;
datalines;
2 1 1
2 2 33
1 1 7
1 2 27
;

proc freq data=predcardiac1 data=order;
tables dose*Response/RELRISK (CL=WALD) ;
weight Count;
run;
```

In PROC FREQ code *RELRISK (CL=WALD)* is specified for computing confidence interval using the Katz et al. method. The keyword in SAS/PROC FREQ is *WALD*. The partial output from SAS is shown in the following.

```
Confidence Limits for the Relative Risk
      Relative Risk = 7.0000
    Type    95% Confidence Limits

    Wald      0.9096        53.8695

      Column 1 (Response = 1)

         Sample Size = 68
```

In the above SAS output, the 95% confidence interval of Katz et al. is (0.9096, 53.8695). The corresponding continuity corrected method can be run using the following SAS code.

```
ods listing;
data predcardiac1;
input Dose Response Count;
datalines;
2 1 1
2 2 33
1 1 7
1 2 27
;

proc freq data=predcardiac1 data=order;
tables dose*Response/RELRISK (CL=WALDMODIFIED) ;
weight Count;
run;
```

Notice that, in PROC FREQ code *RELRISK (CL=WALDMODIFIED)* is specified for computing continuity corrected interval. SAS gives the following partial output.

```
Confidence Limits for the Relative Risk
        Relative Risk = 7.0000
   Modified Relative Risk = 5.0000

Type              95% Confidence Limits

Wald Modified       0.9241        27.0523

       Column 1 (Response = 1)

           Sample Size = 68
```

Thus the continuity corrected 95% interval is $(0.9241, 27.0523)$. Clearly, it is much narrower than the uncorrected interval.

5.2.1.2 Asymptotic score interval: Koopman

Koopman [50] proposed the following asymptotic score interval. Under the null hypothesis $\phi = \phi_0$, Koopman showed the following test statistic follows chi-square distribution with 1 degree of freedom.

$$T_{\phi_0}(\boldsymbol{x}) = \frac{(x_1 - n_1\widetilde{p_1})^2}{n_1\widetilde{p_1}(1-\widetilde{p_1})} + \frac{(x_2 - n_2\widetilde{p_2})^2}{n_2\widetilde{p_2}(1-\widetilde{p_2})} \tag{5.12}$$

where,

$$\widetilde{p_1} = \frac{\phi_0(x_2+n_1) + x_1 + n_2 - \sqrt{\{\phi_0(x_2+n_1)+x_1+n_2\}^2 - 4\phi_0(n_1+n_2)(x_1+x_2)}}{2(n_1+n_2)} \tag{5.13}$$

is the maximum likelihood estimate of p_1 under the restriction $\phi = \phi_0$ and p_2 curl $=$ p_1 curl divided by ϕ_0. Hence the asymptotic two-sided confidence interval of ϕ with confidence level $(1-\alpha)$ is obtained as,

$$\{\phi : T_{\phi}(\boldsymbol{x}) \leq \chi_1^2(\alpha/2)\}, \tag{5.14}$$

where $\chi^2_{\alpha/2}$ is the $(1-\alpha/2)$th percentile of χ^2_1 distribution. Now, $T_\phi(\boldsymbol{x})$ being a convex function of ϕ, one can iteratively compute the two confidence limits $(\underline{\phi}, \overline{\phi})$ as a solution of equation $T_\phi(\boldsymbol{x}) = \chi^2_1(\alpha/2)$.

Using the same setting as in section 5.2.1.1, the simulated coverage probabilities of Koopman score interval are plotted against p_1 in Figure 5.2.

FIGURE 5.2
Coverage probabilities of Koopman score confidence intervals.

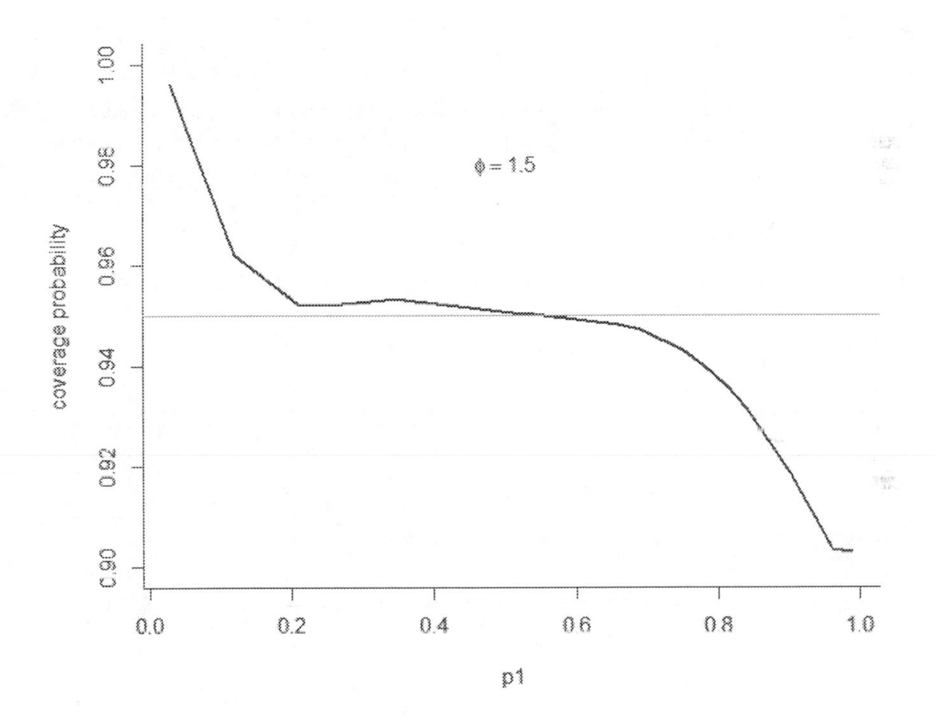

Notice that the Koopman score interval shows a reasonably good coverage (closely aligned with the nominal level) property when p_1 lies in the interval $(0.2, 0.6)$, however, for extreme values of p_1 at either end. However, when the values are towards the extremes of the parameter space $(0, 1)$, the interval is either very conservative or very liberal. The distribution of the simulated coverage probabilities in the $p_1 - p_2$-space is given in the discussion section.

Illustration

For illustration, let us consider the cardiac arrest data given in Table 5.1. We run the following StatXact PROCs code to compute the Koopman interval.

```
data predcardiac;
input Dose Response Count;
datalines;
1 1 1
1 2 33
2 1 7
2 2 27
;
PROC BINOMIAL DATA=predcardiac;
PR/AS ONE NONSTD;
PO Dose;
OU Response;
WEIGHT COUNT;
RUN;
```

In the above code, the high dose is coded Dose=1 and the standard dose is coded Dose=2. The syntax *PR/AS ONE NONSTD* invokes the interval for ratio of proportions (*PR*) using the asymptotic method (*AS*) with non-standardized test statistics (*NONSTD*). StatXact PROCs's PROC BINOMIAL shows the following output:

```
CONFIDENCE INTERVAL ON RATIO OF TWO BINOMIAL PROPORTIONS BASED ON THE UNSTANDARDIZED
STATISTIC AND INVERTING TWO 1-SIDED TESTS

Data file name : < PREDCARDIAC >
Population Variable Name : Dose
Outcome Variable Name : Response
Weight Variable Name : Count


Statistics based on the observed  2 by  2 table :

    Observed proportion for population <       1> : piHat_1          =     0.0294
    Observed proportion for population <       2> : piHat_2          =     0.2059
    Observed ratio of proportions : piHat_2/piHat_1                  =     7.0000
    Stderr (pooled estimate of stdev of piHat_2-piHat_1  )           =     0.0781

                                  (x12+.5)/(n2+.5)
    Ratio adjusted for extremes = ----------------       =       5.0000
                                  (x11+.5)/(n1+.5)
Results:
---------------------------------------------------------------------------
                            P-value            95.00% Conf. Interval
Method                     2*1-sided            for pi_2/pi_1
---------------------------------------------------------------------------
Asymp ( Katz et. al.)        0.0617        (      0.9241,      27.0523)
Asymp ( Koopman )            0.0239        (      1.2209,      42.5772)
```

Thus, the 95% confidence intervals of ϕ using continuity corrected Katz et al. and the Koopman's score interval are $(0.9241, 27.0523)$, and $(1.2209, 42.5772)$, respectively. Note that the Katz et al. interval includes 1 suggesting that at 5% level the null hypothesis $\phi = 1$ cannot be rejected, whereas Koopman's score interval excludes 1, thus rejects it. Fagerland, Lydersen and Laake [34] compared these intervals and concluded that the

Katz et al. interval may be less reliable than Koopman's score interval, and the latter performs better in most cases.

5.2.1.3 Asymptotic score interval: Farrington and Manning

Farrington and Manning [35] proposed the following method to compute the confidence interval of ϕ. Under the null hypothesis $\phi = \phi_0$, the standardized test statistic,

$$S(\boldsymbol{x}) = \frac{\widehat{p_1}/\widehat{p_2} - \phi_0}{\sqrt{\frac{(\widetilde{p}_1)(1-\widetilde{p}_1)}{n_1} + {\phi_0}^2 \frac{(\widetilde{p}_2)(1-\widetilde{p}_2)}{n_2}}} \tag{5.15}$$

has an asymptotic $N(0,1)$ distribution, where, $\widetilde{p}_1$ and $\widetilde{p}_2$ are the constrained maximum likelihood estimates of p_1 and p_2 under the restriction $p_1/p_2 = \phi_0$. The estimates $\widetilde{p}_1$ and $\widetilde{p}_2$ are given by,

$$\begin{aligned} \widetilde{p}_2 &= \frac{-B - \sqrt{B^2 - 4AC}}{2A}, \\ \widetilde{p}_1 &= \phi_0 \widetilde{p}_2, \end{aligned}$$

where

$$\begin{aligned} A &= \phi_0 N, \\ B &= -(\phi_0 n_1 + x_1 + n_2 + \phi_0 x_2), \\ C &= x_1 + x_2, \ N = n_1 + n_2 \end{aligned} \tag{5.16}$$

The asymptotic level $(1-\alpha)$ confidence interval of Farrington and Manning (**FM**) is obtained by inverting two one-sided tests based on $S(\boldsymbol{x})$; which boils down to solving the following two equations:

$$1-\Phi\left(\frac{\widehat{p_1}/\widehat{p_2} - \underline{\phi}}{\sqrt{\frac{(\widetilde{p}_1)(1-\widetilde{p}_1)}{n_1} + {\phi_0}^2 \frac{(\widetilde{p}_2)(1-\widetilde{p}_2)}{n_2}}}\right) = \frac{\alpha}{2} = \Phi\left(\frac{\widehat{p_1}/\widehat{p_2} - \overline{\phi}}{\sqrt{\frac{(\widetilde{p}_1)(1-\widetilde{p}_1)}{n_1} + {\phi_0}^2 \frac{(\widetilde{p}_2)(1-\widetilde{p}_2)}{n_2}}}\right).$$

Considering the same setting as in section 5.2.1.1, the plot of simulated coverage probabilities against p_1 is shown in Figure 5.3.

This interval shows reasonably good coverage (closely aligned with the nominal level) when the true p_1 values are approximately in the interval $(0.2, 0.75)$. However, for the values at both ends of the interval $(0, 1)$, the method is either very conservative or liberal. In this setting, this interval seems to be an improved version of the previously stated Koopman interval, especially in terms of the coverage probabilities for high values of p_1. The distribution of simulated coverage probabilities in the $p_1 - p_2$-space is given in the discussion section.

FIGURE 5.3
Coverage probabilities of Farrington and Manning score confidence intervals.

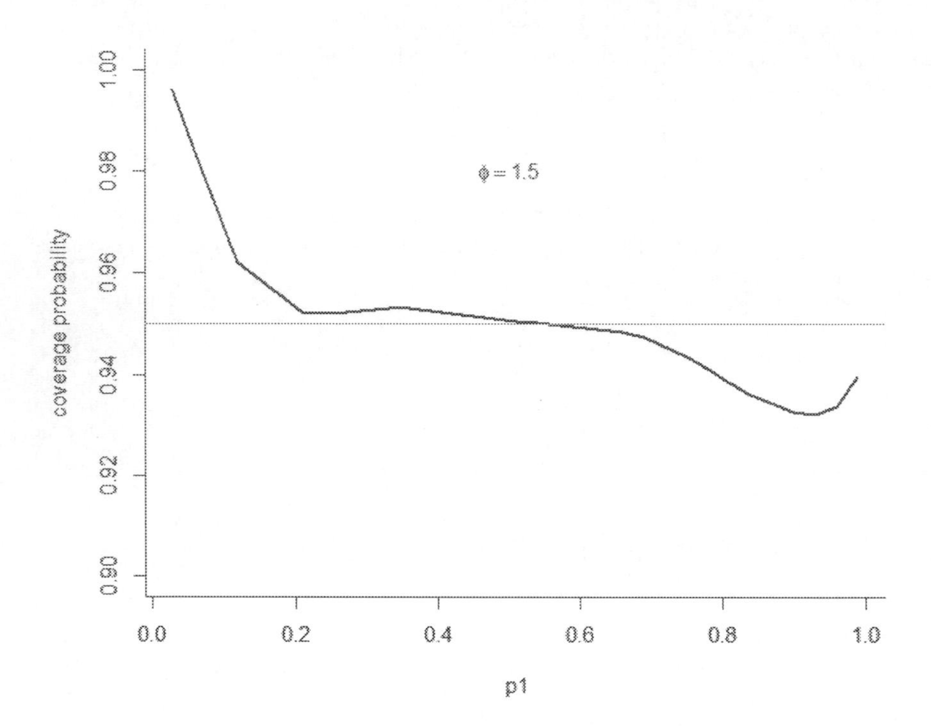

Illustration: Farrington and Manning

For illustration, we use the same cardiac arrest data set. The SAS code using SAS's PROC FREQ is given below for computing the 95% interval of ϕ:

```
ods listing;
data predcardiac1;
input Dose Response Count;
datalines;
2 1 1
2 2 33
1 1 7
1 2 27
;

proc freq data=predcardiac1 data=order;
tables dose*Response/RELRISK (CL=SCORE(CORRECT=NO) ) ;
weight Count;
run;
```

In PROC FREQ's table statement the option *RELRISK (CL=SCORE(CORRECT=NO))* invokes the Farrington and Manning method. Removing the option *RELRISK (CL=SCORE)* enables a correction factor $n/(n-1)$ in the denominator of the test statistic and provides a confidence interval using Miettinen and Nurminen test (described in the next section). Also, in the PROC FREQ statement *data=order* statement is specified to preserve the input data order. SAS's PROC FREQ shows the following partial output.

```
Confidence Limits for the Relative Risk
        Relative Risk = 7.0000
    Type      95% Confidence Limits

    Score      1.2209          42.5757

       Column 1 (Response = 1)

           Sample Size = 68
```

Note that in the output, the *Relative Risk* shows the point estimate $\widehat{\phi} = 7$, and the corresponding 95% confidence interval is reported as $(1.2209, 42.5757)$.

5.2.1.4 Asymptotic score interval: Miettinen and Nurminen

Miettinen and Nurminen [57] (**MN**) proposed a score method for which the standardized test statistic to compute confidence interval is almost identical to that of **FM** method. The key difference is that in computing the variance estimate based on restricted likelihood in the denominator, an additional correction factor $\frac{N}{N-1}$ is multiplied to the variance estimate.

$$S(\boldsymbol{x}) = \frac{\widehat{p_1}/\widehat{p_2} - \phi_0}{\sqrt{\left(\frac{(\widetilde{p}_1)(1-\widetilde{p}_1)}{n_1} + {\phi_0}^2 \frac{(\widetilde{p}_2)(1-\widetilde{p}_2)}{n_2}\right)\left(\frac{N}{N-1}\right)}} \tag{5.17}$$

Finally, the asymptotic level-$(1-\alpha)$ confidence interval is computed by inverting one-sided tests based on $S(\boldsymbol{x})$ similar to **FM** method as shown below:

$$1 - \Phi\left(\frac{\widehat{p_1}/\widehat{p_2} - \underline{\phi}}{\sqrt{\left(\frac{(\widetilde{p}_1)(1-\widetilde{p}_1)}{n_1} + {\phi_0}^2 \frac{(\widetilde{p}_2)(1-\widetilde{p}_2)}{n_2}\right)\left(\frac{N}{N-1}\right)}}\right) = \frac{\alpha}{2}$$

$$= \Phi\left(\frac{\widehat{p_1}/\widehat{p_2} - \overline{\phi}}{\sqrt{\left(\frac{(\widetilde{p}_1)(1-\widetilde{p}_1)}{n_1} + {\phi_0}^2 \frac{(\widetilde{p}_2)(1-\widetilde{p}_2)}{n_2}\right)\left(\frac{N}{N-1}\right)}}\right)$$

Using the same setting as stated in section 5.2.1.1, the simulated coverage probabilities are shown in Figure 5.4.

FIGURE 5.4
Coverage probabilities of Miettinen and Nurminen score confidence intervals.

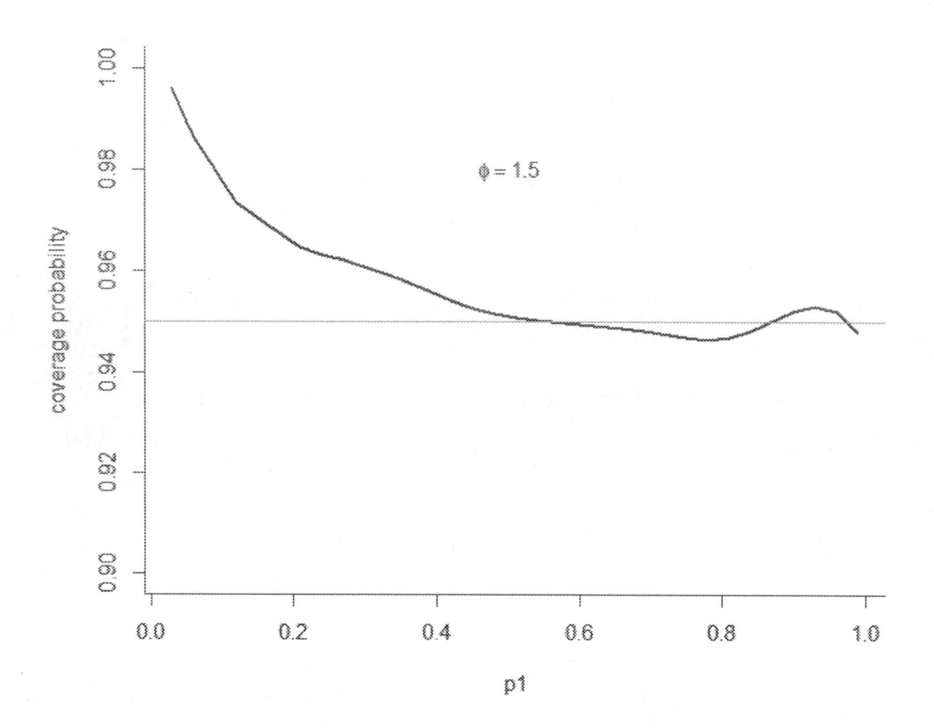

The **MN** interval shows a reasonably good alignment of its coverage probabilities with the nominal level 0.95 for most of the values of p_1, except for values less than 0.2, for which it gives substantial over-coverage. In this setting, the **MN** method appears to be an improved version of the Koopman and the **FM** intervals. Incorporating the correction factor $\frac{N}{N-1}$ in the variance leads to significant improvement of its coverage property. The distribution of coverage probabilities in $p_1 - p_2$ space is given in the discussion section.

Illustration: Miettinen and Nurminen

For illustration of the **MN** interval, we consider an example of a Biosimilar study. Therapeutic equivalence of two large molecules is often tested using the confidence interval approach. Consider the data from the Biosimilar study testing equivalence of the compounds CT-P13 and Remicade, where the latter was considered to be the reference compound. The example can be found in the FDA briefing document for CT-P13 (https://www.fda.gov/downloads/advisorycommittees/committeesmeetingmaterials/drugs/arthritisadvisorycommittee/ ucm484860.pdf). The data presented in Table 5.2 are from a double-blinded randomized trial with RA (rheumatoid arthritis) patients. The primary endpoint was a 20% improvement in RA (ACR20) symptoms at week 30 of the study. The data in Table 5.2 summarizes the results at week 30. Consider that the equivalence margins in terms of ACR20 ratio to be tested are between 0.7 and 1.43.

TABLE 5.2
Biosimilar data testing equivalence between CT-P13 and Remicade using of ACR20 at week 30

CT-P13 3 mg/kg	**Remicade 3 mg/kg**
184/302 (60.9%)	179/304 (58.9%)

We run the following SAS code with the option *relrisk(EQUIVALENCE margin=(0.7, 1.43) method=MN) alpha=0.025*. Notice that the test option *relrisk* invokes the relative risk option. Also, inside the parenthesis *EQUIVALENCE margin=(0.7, 1.43) method=MN* option is for equivalence test with margin specified as (0.7 and 1.43), and the method to be used is specified as Miettinen and Nurminen. Since the hypothesis is to be tested with a one-sided alpha=0.025 level (i.e., with a 95% confidence interval), *alpha=0.025* is specified. Also, with the equivalence option specified, by default SAS computes a 90% confidence interval to test for the equivalence.

```
data Biosimimar;
input Treatment Response Count;
datalines;
2 1 178
2 2 126
1 1 184
1 2 118
;
proc freq data=Biosimimar;
tables Treatment*Response/ relrisk(EQUIVALENCE margin=(0.7, 1.43) method=MN) alpha=0.025;
weight Count;
run;
```

PROC FREQ shows the following output of the above SAS code.

```
            Equivalence Analysis for the Relative Risk

         H0: P1 / P2 <= Lower Margin or >= Upper Margin
          Ha: Lower Margin < P1 / P2 < Upper Margin

 Lower Margin = 0.7    Upper Margin = 1.43    Score (Farrington-Manning) Method
```

```
 Relative Risk     Equivalence Limits      95% Confidence Limits

      1.0406       0.7000      1.4300          0.9127      1.1870
```

It is worth mentioning here, that the above SAS output is obtained using SAS's 9.4 TS1M4. The same code with an earlier version of SAS may show an error message. In the output, the lower margin and the upper margins are shown as 0.7 and 1.43 as specified in the SAS code. Since the computed interval $(0.9127, 1.187)$ is completely within the margin $(0.7, 1.43)$, the null hypothesis is rejected, and hence the hypothesis of equivalence between the test compound (CT-P13) and the reference compound (Remicade) is accepted. Notice that, despite the fact that method=MN is specified in the SAS code, the output is shown for *Score (Farrington-Manning)*. As stated before, the test statistic of MN and FM are almost the same, and the only difference comes from an additional correction factor $(N/N-1)$ multiplied to the variance in the denominator. In most situations with moderately large samples, both methods would show numerically identical results.

We can also use PROC BINOMIAL of SatXact PROCs software to get the same interval using the MN method. As before, the syntax of PROC BINOMIAL is almost the same, here the option *PR/AS ONE STD* invokes the MN method.

```
data Biosimimar;
input Treatment Response Count;
datalines;
1 1 178
2 1 184
1 2 126
2 2 118
;
PROC BINOMIAL DATA=BIOSIMIMAR;
PR/AS ONE STD;
PO TREATMENT;
OU RESPONSE;
WEIGHT COUNT;
RUN;
```

```
CONFIDENCE INTERVAL ON RATIO OF TWO BINOMIAL PROPORTIONS BASED ON THE STANDARDIZED
 STATISTIC AND INVERTING TWO 1-SIDED TESTS

Data file name : < BIOSIMIMAR >
Population Variable Name : Treatment
Outcome Variable Name : Response
Weight Variable Name : Count


Statistics based on the observed  2 by  2 table :

    Observed proportion for population <       1> : piHat_1          =      0.5855
    Observed proportion for population <       2> : piHat_2          =      0.6093
    Observed ratio of proportions : piHat_2/piHat_1                  =      1.0406
    Stderr (pooled estimate of stdev of piHat_2-piHat_1  )           =      0.0398
    Standardized difference (t): (piHat_2-piHat_1)/Stderr            =      0.5959
Results:
-------------------------------------------------------------------------------
```

```
                          P-value                             95.00% Conf. Interval
Method          1-sided(Pr{T .GE. t})    2*1-sided                for pi_2/pi_1
-----------------------------------------------------------------------------------
Asymp                  0.2756                0.5512        (      0.9127,       1.1870)
```

Note that the confidence limits are identical to limits using SAS's PROC FREQ method.

5.2.1.5 Profile likelihood interval

This section presents the method of computing confidence interval (Cf. Section 4.2.2.4) based on profile likelihood. As stated before, for Table 4.1, the likelihood can be written as

$$L(\boldsymbol{x}:p_1,p_2)=\prod_{j=1}^{2}\binom{n_j}{x_j}p_j^{x_j}(1-p_j)^{n_j-x_j}$$

where p_i is the probability of success for two population i, $i=1,2$; $0\leq x_1\leq n_1$ and $0\leq x_2\leq n_2$. Since $\phi=p_1/p_2$, and $p_1\neq 0$ and $p_2\neq 0$, the likelihood can be written in terms of ϕ and a nuisance parameter p_1 as shown below:

$$L(\boldsymbol{x}:p_1,\phi)=\prod_{j=1}^{2}\binom{n_j}{x_j}p_1^{x_1}(1-p_1)^{n_1-x_1}(p_1/\phi)^{x_2}(1-p_1/\phi)^{n_2-x_2} \tag{5.18}$$

Now, similar to Agresti and Caffo [1] 's adjustment by adding a non-negative value a to each cell of the observed table, the kernel of the log-likelihood becomes:

$$\begin{aligned} l(p_1,\phi) &= ln\left[L(p_1,\phi)\right] \\ &\propto (x_1+a)ln(p_1)+(n_1-x_1+a)ln(1-p_1)+(x_2+a)ln(p_1/\phi) \\ &\quad +(n_2-x_2+a)ln(1-p_1/\phi) \end{aligned} \tag{5.19}$$

where $p_1\neq 0$ and $\phi\neq 0$. Following Venzon and Moolgavkar (1998), the approximate $1-\alpha$ confidence interval based on profile likelihood method for ϕ is

$$\{\phi:2[l(\widehat{\phi},\widehat{p}_1)-l(\phi,\widetilde{p}_1)]\leq\chi_1^2(\alpha)\} \tag{5.20}$$

where $\widehat{\phi}$ and $\widehat{p}_1$ are the unrestricted maximum likelihood estimates of ϕ and p_1, $\widetilde{p}_1$ is the restricted maximum likelihood of p_1 for given ϕ, and $\chi_1^2(\alpha)$ is the upper 100α percentile point of the chi-square distribution with 1 degree of freedom. Venzon and Moolgavkar (1988) then suggested, the confidence interval for ϕ, is the admissible solution to the following system of non-linear equations:

$$\begin{bmatrix} l(\widehat{\phi},\widehat{p}_1) - l(\phi,p_1) - \frac{1}{2}\chi_1^2(\alpha) \\ \frac{\partial l(\phi,p_1)}{\partial p_1} \end{bmatrix} = 0.$$

The SAS/IML implementation of the profile likelihood is straightforward and similar to the previously given implementation of the difference of proportions. The following SAS/IML can be used to compute the asymptotic profile likelihood-based confidence interval. It should be noted that because of the computational issues, a very high number as an upper limit can be treated as infinity.

```
%macro p1p2_ratio_profile(sample1= ,sample2= ,success1= ,success2= , alpha= , a= );
%if &success1.=0 %then %let success1=%SYSEVALF(&success1.+0.005);
%if &success2.=0 %then %let success2=%SYSEVALF(&success2.+0.005);
%let s1=%SYSEVALF(&success1);
%let s2=%SYSEVALF(&success2);
%let n1=%SYSEVALF(&sample1);
%let n2=%SYSEVALF(&sample2);
proc iml;
Start likelihood(x)global(n1,n2,s1,s2) ;
      sum1=0.0;sum2=0.0;sum3=0.0;sum4=0.0;

      if &a.=0.5 then do;
      a=0.5;
      end;
      else a=n2/(n1+n2);

      if x[1]#x[2]<=0 then demo1=1e-4;
      else if x[1]#x[2]>=1 then demo1=0.9999;
      else demo1=x[1]#x[2];

      if x[2]<=0 then demo2=1e-4;
      else if x[2]>=1 then demo2=0.9999;
      else demo2=x[2];

      sum1=(s1+a)#log(demo1);
      sum2=(n1-s1+a)#log(1-demo1);
      sum3=(s2+1-a)#log(demo2);
      sum4=(n2-s2+1-a)#log(1-demo2);
      f=sum1+sum2+sum3+sum4;
      return (f);
finish likelihood;

start Gradient(x)global(n1,n2,s1,s2);
      g=j(1,2,1e-4);

      if &a.=0.5 then do;
       a=0.5;
      end;
      else a=n2/(n1+n2);

      if x[1]#x[2]<=0 then demo1=1e-4;
      else if x[1]#x[2]>=1 then demo1=0.9999;
      else demo1=x[1]#x[2];

      if x[2]<=0 then demo2=1e-4;
      else if x[2]>=1 then demo2=0.9999;
      else demo2=x[2];

      g[1]=(s1+a)#x[2]/(demo1) - (n1-s1+a)#x[2]/(1.0-demo1);
      g[2]=(s1+a)#x[1]/(demo1) - (n1-s1+a)#x[1]/(1.0-demo1) + (s2+1-a)/demo2 - (n2-s2+1-a)/
      (1.0-demo2);
      return (g);
finish Gradient;
```

```
 n = 2; n1=&n1. ;n2=&n2. ;s1=&s1. ;s2=&s2.;

 x0 ={1e-9 0.999999999};
/* x0 ={-1 1};*/
 optn = {1 0};
 con = { 1.e-9 1.e-9 . .,
              999999999 0.999999999 . . ,
             1 1 1 1e-9,
             1 1 -1 999999999};
 call nlptr(rc,xres,"likelihood",x0,optn,con,,,,"Gradient");
 xopt = xres`; fopt = likelihood(xopt);

 /* for Hessian */
 call nlpfdd(f,g,hes2,"likelihood",xopt,,"Gradient");

 start plgrad(x) global(like,ipar,lstar);
 like = likelihood(x);
 grad = Gradient(x);
 grad[ipar]=like-lstar;
       return(grad`);
 finish plgrad;

      prob=&alpha ;
       xlb=j(2,1,0);
      xub=j(2,1,9999);
      /* quantile of chi**2 distribution */
      chqua = cinv(1-prob,1);like=fopt; lstar = fopt - .5 * chqua;

 optn = {2 0};
 do ipar = 1 to 2;
       /* Implementation of Venzon & Moolgavkar (1988)*/
       if ipar=1 then ind = 2; else ind = 1;
       delt = - inv(hes2[ind,ind]) * hes2[ind,ipar];
       alfa = - (hes2[ipar,ipar] - delt` * hes2[ind,ipar]);
       if alfa > 0 then alfa = .5 * sqrt(chqua / alfa);
       else do;
       print "Bad alpha";
       alfa = .1 * xopt[ipar];
       end;
       if ipar=1 then delt = 1 || delt;
       else delt = delt || 1;
       x0 = xopt + (alfa * delt)`;
       con2 = con; con2[1,ipar] = xopt[ipar];
             tc={10000,10000};
       call nlplm(rc,xres,"plgrad",x0,optn,con2,tc );

       f = plgrad(xres); s = ssq(f);

             if (s <1.e-4) then xub[ipar] = xres[ipar];
                         else xub[ipar] =9999;
             x0 = xopt - (alfa * delt)`;
       con2[1,ipar] = con[1,ipar]; con2[2,ipar] = xopt[ipar];
             tc={10000,10000};
       call nlplm(rc,xres,"plgrad",x0,optn,con2,tc);
       f = plgrad(xres); s = ssq(f);
       if (s < 1.e-4) then xlb[ipar] = xres[ipar];
       else xlb[ipar] = 0.0;

 end;

ci=xlb||xub;

create Profile from ci[colname={'Lower','Upper'} ] ;
append from ci;
close Profile;
run;
```

```
quit;

data Profile; set Profile;
if _n_=2 then delete;
run;

Proc print data=Profile noobs label;
title1 "*****************************************************************************";
title2 "* Confidence interval of ratio of two binomial proportions *";
title3 "* Based on Weighted Profile Likelihood Method   *";
title5 "*****************************************************************************";
run;
%mend;
```

Using the same setting as in section 5.2.1.1, the plot of coverage probability values against p_1 is given in Figure 5.5. The weighted profile likelihood interval shows good coverage properties except for values of p_1 less than 0.2 like the previous intervals. In general, the weighted profile likelihood appears to be a reasonably good asymptotic method, and better than other asymptotic methods introduced so far in this chapter. Interestingly, in this setting, it never gives under-coverage. The distribution of the coverage probabilities in $p_1 - p_2$ space is given in the discussion section.

Illustration: Profile likelihood interval

Let us compute the 95% confidence interval for the data introduced in Table 5.1. We run the above SAS macro using the following code:

```
%p1p2_ratio_profile(sample1=34 ,sample2=34 ,success1=7 ,success2=1 , alpha=0.05 , a= 0.5);
```

SAS shows the following output:

Lower	Upper
1.18926	45.6468

Notice that the confidence interval $(1.18926, 45.6468)$ excludes 1, which is consistent with the other score intervals. Also, note that on some rare occasions, the above code may produce erroneous results. In those situations, one can apply the property given in equation 5.1 to get a confidence interval indirectly. For example, one could run the following code by interchanging the populations and the corresponding successes.

```
%p1p2_ratio_profile(sample1=34 ,sample2=34 ,success1=1 ,success2=7 , alpha=0.05 , a= 0.5);
```

SAS shows the following output:

Lower	Upper
0.02191	0.84086

FIGURE 5.5
Coverage probabilities of weighted profile likelihood confidence intervals.

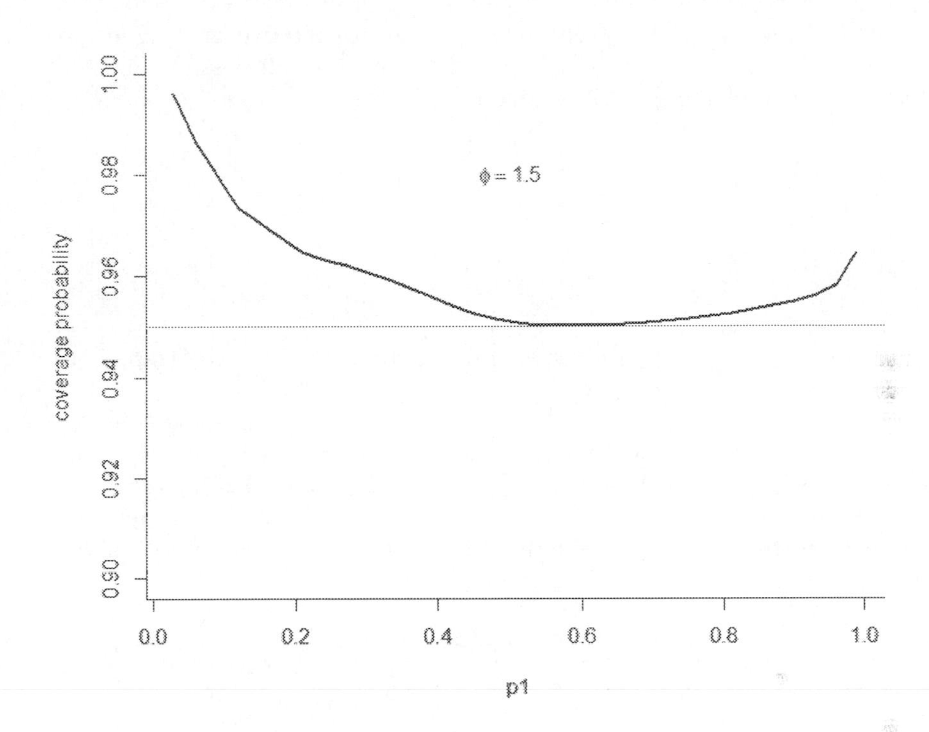

Now, using equation 5.1 one can compute the lower limit as $1/0.84086 = 1.189258616$ and the upper limit as $1/0.02191 = 45.64750993$ (since, if $(\underline{\phi}, \overline{\phi})$ is the confidence interval of p_1/p_2, then the confidence interval of p_2/p_1 is $(1/\overline{\phi}, 1/\underline{\phi})$), which is same as $(1.18926, 45.6468)$ that we computed earlier using the code

```
%p1p2_ratio_profile(sample1=34 ,sample2=34 ,success1=7 ,success2=1 , alpha=0.05 , a= 0.5);
```

5.2.2 Exact intervals

In this section, we introduce two important exact confidence intervals for the ratio of two binomial proportions. The methods rely on the exact distribution to derive the confidence interval. Like other exact methods, discussed in this book, these intervals are also computationally challenging, but they attain the nominal coverage level.

5.2.2.1 Chan and Zhang interval

The Chan and Zhang interval ([22]) for the proportion is obtained by mimicking the approach they adopted for finding a confidence interval for the difference of two binomial proportions discussed in section 4.2.3.1.

Let $(\underline{\phi}^{CZ}, \overline{\phi}^{CZ})$ be the level-$(1-\alpha)$ confidence interval of ϕ. Then for $\boldsymbol{\Omega} = \{\boldsymbol{x} = (x_1, x_2) : 0 \leq x_j \leq n_j, j = 1, 2\}$, the set of all possible binomial outcomes, the probability of observing $\boldsymbol{x}$ is

$$f(\boldsymbol{x}) = f(\boldsymbol{x}|p_1, \phi) = \prod_{j=1}^{2} \binom{n_j}{x_j} p_j^{x_j} (1-p_j)^{n_j - x_j}$$

where, $p_2 = \phi p_1$. Let

$$P_{p_1,\phi}(S(\boldsymbol{x})) = \sum_{\boldsymbol{y}:S(\boldsymbol{y})\leq S(\boldsymbol{x})} f(\boldsymbol{y}|p_1, \phi) \quad \left(Q_{p_1,\phi}(S(\boldsymbol{x})) = \sum_{\boldsymbol{y}:S(\boldsymbol{y})\geq S(\boldsymbol{x})} f(\boldsymbol{y}|p_1, \phi) \right) \tag{5.21}$$

denote the probability of obtaining a value of the score statistic that is less than or equal (greater than or equal) to $S(\boldsymbol{x})$. In Equation (4.28), p_1 is viewed as a nuisance parameter with $p_1 \in I(\phi) \equiv (0, \min(1/\phi, 1))$. The nuisance parameter is eliminated by considering the worst-case p_1 scenario of obtaining small (large) values of $S(\boldsymbol{x})$:

$$P_\phi(S(\boldsymbol{x})) = \sup\{P_{p_1,\phi}(S(\boldsymbol{x})) : p_1 \in I(\phi)\}$$

and

$$Q_\phi(S(\boldsymbol{x})) = \sup\{Q_{p_1,\phi}(S(\boldsymbol{x})) : p_1 \in I(\phi)\}$$

The level $100(1-\alpha)\%$ **CZ** confidence interval $(\underline{\phi}^{CZ}, \overline{\phi}^{CZ})$ is the solution of

$$P_{\overline{\phi}^{CZ}}(S(\boldsymbol{x})) = \frac{\alpha}{2} = Q_{\underline{\phi}^{CZ}}(S(\boldsymbol{x})).$$

In the same setting as in section 5.2.1.1, the simulated coverage probabilities against p_1 are shown in Figure 5.6.

It is evident from the Figure 5.6 that the Chan and Zhang interval guarantees the nominal coverage level, and is conservative (wider confidence intervals). The distribution of the coverage probabilities in the $p_1 - p_2$ space is discussed at the end.

Illustration: Chan and Zhang interval

For illustration, we consider the vaccine efficacy data discussed in Chan [21] and reported in Fries et al. [38]. The data relate to a randomized trial for the

FIGURE 5.6
Coverage probabilities of Chan and Zhang confidence intervals.

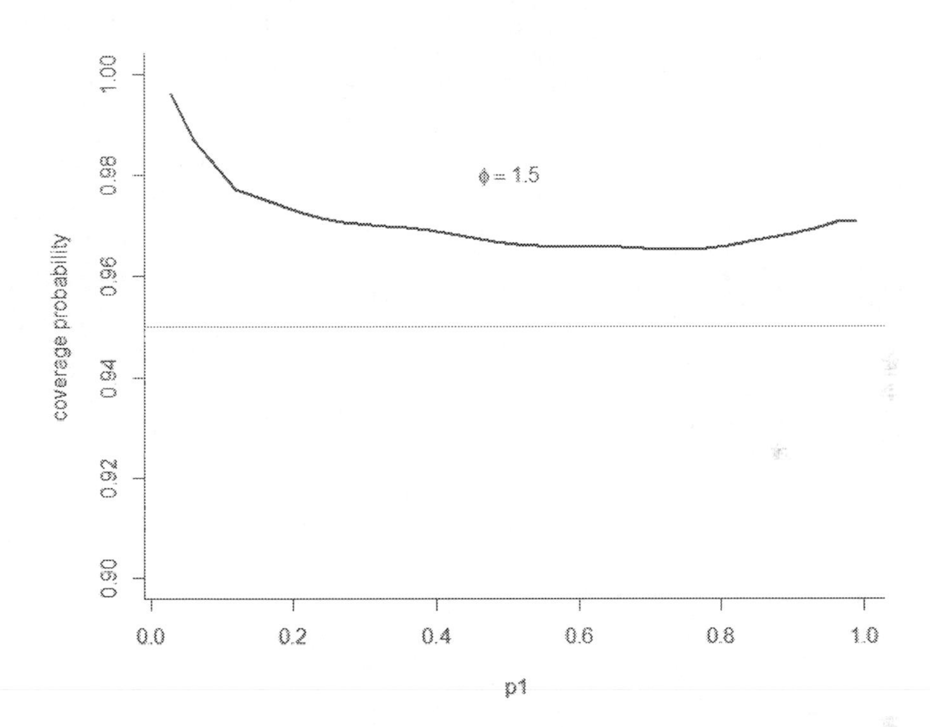

study of efficacy of a Recombinant protein Influenza **A** vaccine against wild-type H1N1 virus. In this study, 15 subjects were given the influenza vaccine and 15 were given a placebo vaccine. The incident rates of the viral infection at day 9 were compared. The results are given in Table 5.3.

Let p_1 and p_2 be the infection rates of the placebo and influenza vaccines, respectively. Then the observed infection rate, say $\widehat{\phi}$, is equal to $0.47/0.8 = 0.5875$. Let $\pi = 1 - \phi$ be the efficacy rate. Notice that if the observed infection rate due to influenza vaccine is $\widehat{p}_2 = 0$, then $\widehat{\phi} = 0$, and $\widehat{\pi} = 1$ implying 100% estimated efficacy of the vaccine. Suppose the influenza vaccine would be acceptable if the efficacy rate is greater than 10%. Chan [21] considered the hypothesis $H_0 : \pi = 0.1$ versus $H_1 : \pi > 0.1$ with $\alpha = 0.025$. Since $\pi = 1 - \phi$, one can equivalently test,

$$H_0 : \phi \geq 0.9 \quad \text{versus} \quad H_1 : \phi < 0.9.$$

To test the one-sided hypothesis at $\alpha = 0.025$ level, we compute a 2-sided 95%

TABLE 5.3
Vaccine efficacy study data against wild-type H1N1 virus

	Placebo	**Vaccine**
Infected	12 (80%)	7 (47%)
Non-infected	3 (20%)	8 (53%)
Total	15 (100%)	15 (100%)

confidence interval. The SAS code for the 95% exact confidence interval due to Chan and Zhan [22] is given below:

```
ods html close;
ods listing;
data h1n1vaccine;
input Treatment Response Count;
datalines;
1 1 7
1 2 8
2 1 12
2 2 3
;
proc freq data=h1n1vaccine;
tables Treatment*Response;
exact relrisk (method=score);
weight Count;
run;
```

The following is the SAS output:

```
                    Relative Risk (Column 1)
----------------------------------------------------------
          Relative Risk                    0.5833

          Asymptotic Conf Limits
          95% Lower Conf Limit             0.3210
          95% Upper Conf Limit             1.0600

          Exact Conf Limits
          95% Lower Conf Limit             0.2608
          95% Upper Conf Limit             1.0372

     The exact confidence limits are based on the score statistic.

                         Sample Size = 30
```

Thus, the exact 95% confidence interval is reported as $(0.2608, 1.0372)$. Since the upper confidence limit 1.0372 is greater than 0.9, we cannot reject the null hypothesis at level 0.025.

One can also test the same hypothesis using StatXact/StatXact PROCs PROC BINOMIAL. Since PROC BINOMIAL by default computes the confidence interval of p_2/p_1, one has to adjust the data set accordingly. Since the hypothesis is equivalent to testing the non-inferiority with margin 0.9, the following syntax can be used in StatXact PROCs:

```
data h1n1vaccine1;
```

```
input Treatment Response Count;
datalines;
1 1 12
1 2 3
2 1 7
2 2 8
;
PROC BINOMIAL DATA=h1n1vaccine1 gamma=0;
NONINF/EX  RATIO MARGIN=0.9;
PO Treatment;
OU Response;
WEIGHT COUNT;
RUN;
```

In the above code, *NONINF/EX RATIO MARGIN=0.9* is for non-inferiority test using exact (*EX*) method for the *RATIO* with a *MARGIN=0.9*. StatXact PROCs shows the following output:

```
UNCONDITIONAL TEST OF NON-INFERIORITY USING RATIO OF TWO BINOMIAL PROPORTIONS

Data file name : < H1N1VACCINE1 >
Population Variable Name : Treatment
Outcome Variable Name : Response
Weight Variable Name : Count

    H0:(pi_2/pi_1) .GE. rho_0 vs H1: (pi_2/pi_1) .LT. rho_0

Statistic based on the observed  2 by  2 table :

    Observed proportion for population <         1> : piHat_1              =        0.8000
    Observed proportion for population <         2> : piHat_2              =        0.4667
    Observed ratio of proportions : piHat_2/piHat_1                        =        0.5833
    Maximum  margin of non-inferiority  : pi_2/pi_1 = rho_0                =        0.9000
    Stderr(restricted mle of stdev of piHat_2-piHat_1*rho_0 given rho_0)   =        0.1689
    Standardized test statistic: (piHat_2-piHat_1*rho_0)/Stderr            =       -1.4999

Results:
----------------------------------------------------------------------
                1-sided P-value                 97.50% Upper Confidence
     Method       Pr{T .LE. t}                    Bound for pi_2/pi_1
----------------------------------------------------------------------
      Asymp           0.0668                          1.0294
      Exact           0.0857                          1.0372
```

Note that the asymptotic 97.5% upper bound is different from that of given by SAS's PROC FREQ output since the SAS PROC FREQ computes the Wald interval whereas the StatXact PROCs computes the Miettinen and Nurminen interval. But, the upper limit 1.0372 matches with the PROC FREQ's upper limit. Besides computing the one-sided 97.5% upper bound, StatXact also computes the asymptotic and unconditional exact p-values. We refer the interested readers to the StatXact PROCs user manual for further details.

5.2.2.2 Agresti and Min interval

Similar to finding Agresti and Min interval for the difference of two binomial proportions in section 4.2.3.2, an exact confidence interval for the ratio of binomial proportion can be computed as follows. Let $(\underline{\phi}^{AM}, \overline{\phi}^{AM})$ be the level $(1-\alpha))$ confidence interval of ϕ. Set

$$R_{p_1,\phi}(S(\boldsymbol{x})) = \sum_{\{\boldsymbol{y}:|\ S(\boldsymbol{y})|\leq|S(\boldsymbol{x})|\}} f_{p_1,\phi}(\boldsymbol{y}).$$

The nuisance parameter p_1 is then eliminated by taking the supremum of $R_{p_1,\phi}$ over values of p_1 in the interval $I(\phi) \equiv (0, \min(1/\phi, 1))$. We set

$$R_\phi(S(\boldsymbol{x})) = \sup\{R_{p_1,\phi}(S(\boldsymbol{x})) : p_1 \in I(\phi)\}.$$

Then the level $(1-\alpha)$ **AM** interval $(\underline{\phi}_{AM}, \overline{\phi}^{AM})$ is obtained by setting $\underline{\phi}_{AM}$ as the solution to

$$R_{\underline{\phi}_{AM}}(S(\boldsymbol{x})) = \alpha,$$

and $\overline{\phi}^{AM}$ a solution to

$$R_{\overline{\phi}^{AM}}(S(\boldsymbol{x})) = \alpha.$$

Using the same setting as in section 5.2.1.1, the coverage probability distribution over p_1 is shown in the Figure 5.7.

It is evident from the Figure 5.7, the Agresti and Min interval attains the nominal coverage level for all values of p_1. However, the interval is conservative (wider intervals) leading to over-coverage, especially for values closer to $p_1 = 0$. In terms of the attainment of the nominal coverage probability, the Agresti and Min interval appears to be slightly better than the Chan and Zhang interval. The distribution of its coverage probabilities in the $p_1 - p_2$ space is taken up in the discussion section.

Illustration: Agresti and Min interval

To illustrate the Agresti and Min method, we consider the data presented in Table 5.1. To the best of our knowledge, the **AM** interval for the ratio of proportions is implemented only in StatXact PROCs software and PROC FREQ (SAS 9.4). The following code can be used to compute the 95% confidence interval:

```
data predcardiac2;
input Dose Response Count;
datalines;
1 1 1
1 2 33
2 1 7
2 2 27
;

PROC BINOMIAL DATA=predcardiac2 gamma=0;
PR/EX TWO STD;
PO Dose;
OU Response;
WEIGHT COUNT;
RUN;
```

In the above code *PR/EX TWO STD* is for the ratio of proportions (*PR*) test using exact (*EX*) method by inverting a two-sided (*TWO*) standardized test statistic (*STD*). StatXact PROCs shows the following output:

FIGURE 5.7
Coverage probabilities of Agresti and Min confidence intervals.

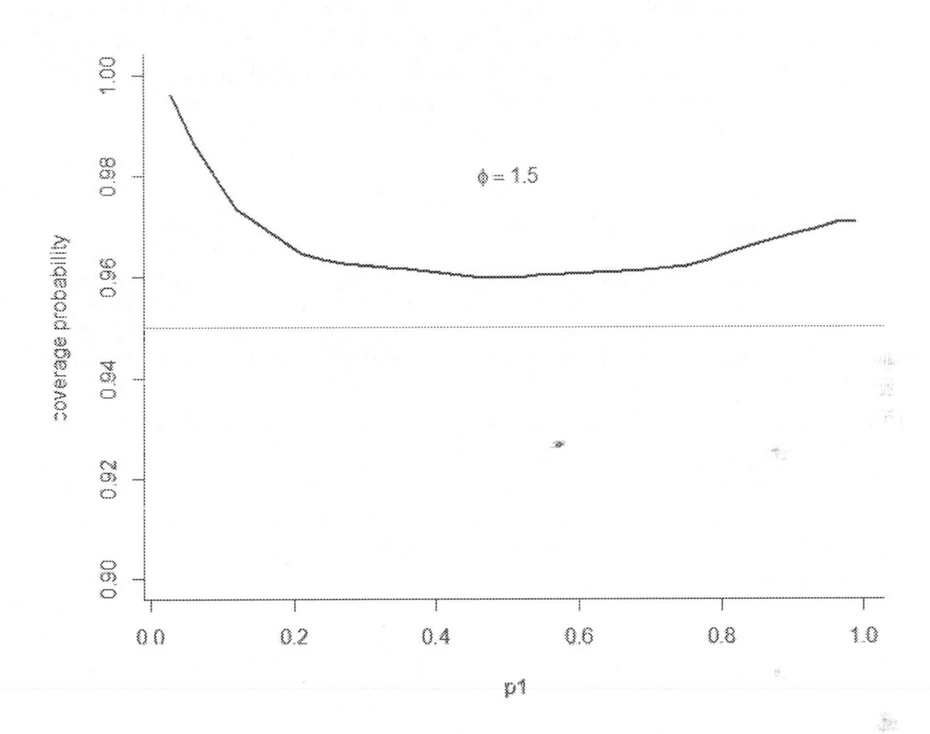


```
CONFIDENCE INTERVAL ON RATIO OF TWO BINOMIAL PROPORTIONS BASED ON THE STANDARDIZED
 STATISTIC AND INVERTING A 2-SIDED TEST

Data file name : < PREDCARDIAC2 >
Population Variable Name : Dose
Outcome Variable Name : Response
Weight Variable Name : Count



Statistics based on the observed  2 by  2 table :

    Observed proportion for population <        1> : piHat_1          =      0.0294
    Observed proportion for population <        2> : piHat_2          =      0.2059
    Observed ratio of proportions : piHat_2/piHat_1                   =      7.0000
    Stderr (pooled estimate of stdev of piHat_2-piHat_1  )            =      0.0781
    Standardized difference (t): (piHat_2-piHat_1)/Stderr             =      2.2583

Results:
---------------------------------------------------------------------------------
                                 P-value                      95.00% Conf. Interval
Method                 2-sided(Pr{|T| .GE. |t|})                  for pi_2/pi_1
```

```
--------------------------------------------------------------------------------
Asymp                    0.0239                    (        1.2086,     43.0331)
Exact                    0.0281                    (        1.1482,     90.3643)
```

In the above, the **MN** interval (asymptotic), and the **AM** interval (exact) are computed. Notice that both the intervals exclude 1, thus, indicating the statistical significance of the conclusion that the standard dose is better than the higher dose for 24-hour survival from cardiac arrest.

5.3 Bayesian intervals

In this section, we briefly discuss the Bayesian computation of the credible intervals of the ratio of two independent binomial proportions (ϕ).

The computation of the interval is very similar to that discussed in Section 4.3. Let's consider the TAXUS ATLAS trial [92] data introduced in section 4.2.2.1 again. Here, the parameter of interest is $h(\boldsymbol{\theta}) = \phi = p_1/p_2$. Let us assume a uniform $(0, 1)$ prior (a non-informative prior) for each of the parameters p_1 and p_2. After a slight modification of the code given in section 4.3, we run the following:

```
data one;
input trt y total;
datalines;
1 67 956
2 68 855
;
ods graphics on;
proc mcmc data=one seed=12346 nmc=100000 nbi=500 outpost=out1 nthin=10 monitor=(phi);
   array p[2] p1 p2;
   parms p1 p2;
   prior p1 p2 ~uniform(0,1);
   model y ~ binomial(total, p[trt]);
   beginnodata;
   phi=p1/p2;
   endnodata;
run;
ods graphics off;
```

The diagnostic plots shown in Figure 5.8 from PROC MCMC indicates no issue with the Markov chains, and the posterior summary for the ratio of proportions ϕ shows that the 95% HPD interval is $(0.6011, 1.1774)$.

Posterior Summaries and Intervals

Parameter	N	Mean	Standard Deviation	95% HPD Interval	
phi	10000	0.8919	0.1493	0.6011	1.1774

FIGURE 5.8
Bayesian diagnostics for the ratio of proportion from PROC MCMC.

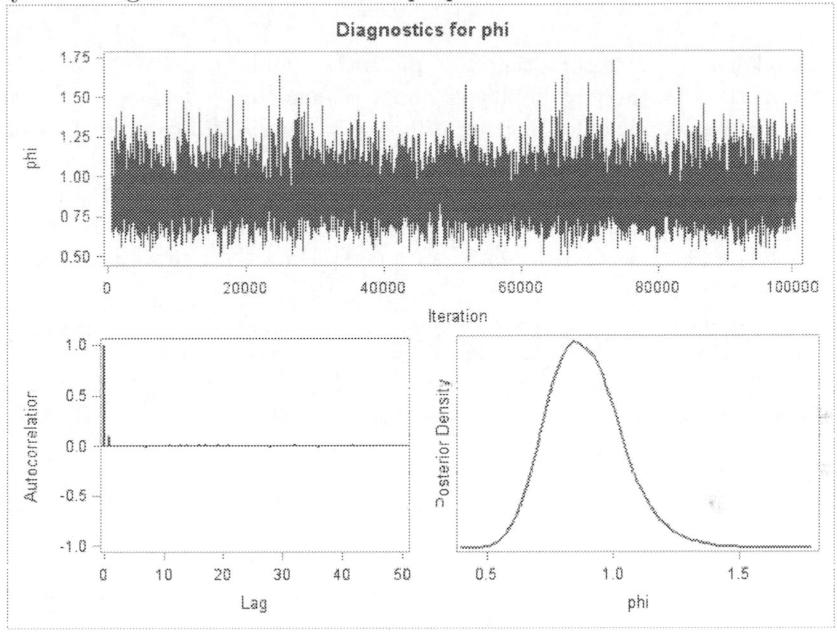

5.4 Discussion and recommendation

In this chapter, we have discussed several commonly used confidence intervals for the ratio of two independent binomial proportions. For large sample sizes, all these intervals should work well. However, in real-world applications, especially for drawing inference from clinical trial data, the most important criterion is to choose an interval that guarantees to attain the nominal level even for moderate sample sizes. All the exact intervals discussed above, by virtue of its method of construction, such as Chan and Zhang [22], Agresti and Min [2], Santner and Snell (PROC FREQ 'S default option to compute a confidence interval for relative risk), guarantee it for any sample sizes. However, for moderately large sample sizes, some of these methods are computationally challenging. In addition, these intervals are often conservative (much wider than necessary). Asymptotic methods, on the other hand, are easy to compute but often fail to attain the nominal coverage probability. Andres and Hernandez [56] evaluated 73 intervals for ϕ, and found none of the traditional intervals is acceptable since all are too liberal. They recommended the modified score interval (after adding 0.5 to all the data points) considering its performance and computational simplicity. Dann and Koch [31] also reviewed some methods of computing the risk ratios. Recently, Fagerland et al. [34] gave an excellent review where most of these intervals are discussed. Fagerland, et al. [34] have preferred the asymptotic score method due to Koopman.

In the following, we present the BliP plots (Lee and Tu [53]) of the distributions of the coverage probabilities and the expected lengths of all of the intervals discussed above for varying values of p_1 and p_2. Since the values of ϕ range between 0 to ∞, the expected length is computed by log-transforming the length and allowing the upper confidence limit to go up to 999. For a given (p_1, p_2) the coverage probability $(\zeta(p_1, p_2))$ and the expected length $(\xi(p_1, p_2))$ of the intervals for $n_1 = n_2 = 20$ are computed using the following formulae,

$$\zeta(p_1, p_2) = \sum_{\boldsymbol{x} \in \Gamma} \prod_{j=1}^{2} \binom{n_j}{x_j} p_j^{x_j} (1 - p_j)^{n_j - x_j} \tag{5.22}$$

where $\Gamma = \{\boldsymbol{x} = (x_1, x_2) : \underline{\phi}(\boldsymbol{x}) \leq p_1/p_2 \leq \overline{\phi}(\boldsymbol{x})\}$, and

$$\xi(p_1, p_2) = \sum_{\boldsymbol{x}} \prod_{j=1}^{2} \binom{n_j}{x_j} p_j^{x_j} (1 - p_j)^{n_j - x_j} \lambda(\boldsymbol{x}), \tag{5.23}$$

where, $\lambda(\boldsymbol{x}) = log(\overline{\phi}(\boldsymbol{x}) - \underline{\phi}(\boldsymbol{x}))$ is the length of the computed confidence interval. Like previous chapters, the unit square $\{(p_1, p_2) | 0 \leq p_1, p_2 \leq 1\}$ is divided into $1000 \times 1000 = 1000^2$ grids. Next, for each grid point, the coverage and expected lengths are computed using the formulas stated in equations 5.22 and 5.23, respectively. As before, the vertical bars in each plot represent the deciles of the corresponding distribution, and the long vertical line in the coverage plot is for the 95% coverage level. For $n_1 = 20$ and $n_2 = 20$, the coverage probability distribution of all asymptotic and exact intervals are shown in Figure 5.9.

Notice that for **KZC** interval (Continuity corrected **KZ** interval), five vertical bars are located to the left of the 95% level, thus, indicating that at least 50% of the grid points fail to achieve the nominal coverage level, and the lowest coverage is around 35%. However, the 10th percentile (the first vertical bar from left to right) is at around 92%, thus, for 10% grid points attained coverage is less than 92% level. The unusually long Blip plot of the **KZC** interval compared to others makes the visual comparisons difficult. Clearly, the **KZC** interval performs the worst, which is consistent with the earlier findings (Price and Bonett [74]).

To facilitate direct visual comparisons, in Figure 5.10, we present the plots of intervals excluding the **KZC** method. From the Figure 5.10, it is evident that among the asymptotic intervals, **WPF** and **MN** appear to perform the best, as only for nearly 20% of the grid points these intervals fail to achieve the nominal coverage level. Both the exact intervals, viz., **CZ** and **AM**, achieve the nominal level. However, the length of the **AM** interval is shorter than the

FIGURE 5.9
Distribution of the coverage probabilities of 95% confidence intervals for $n_1 = 20$ and $n_2 = 20$. KZC=Katz et al. continuity corrected, KPM=Koopman, FM=Farrington and Manning MN=Miettinen and Nurminen WPF=Weighted profile likelihood, CZ=Chan and Zhang, AM=Agresti and Min.

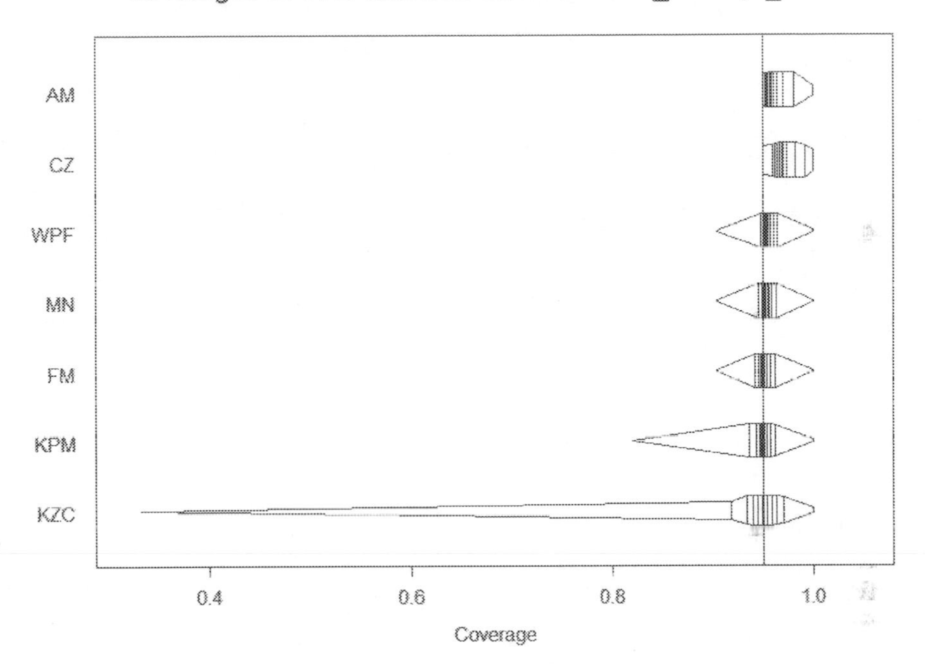

CZ interval since most of the vertical bars in its Blip plot are closely aligned towards the 95% levels.

The Figure 5.11 presents the expected length distributions of all the seven intervals. The intervals **MN** and **FM** appear to be shorter compared to others. The interval **WPF** appears to have a very few grid points where the expected lengths are high.

In section 5.2.2.1, we have discussed the application of Chan and Zhang (**CZ**) interval for testing the non-inferiority of the influenza vaccine with reference to a placebo vaccine. Here, we present the Blip plots of the distribution of the coverage probability for testing the non-inferiority using the same sample sizes, viz., $n_1 = n_2 = 34$. For drawing the Blip plots shown in Figure 5.12, the parameter space of $p = (p_1, p_2)$ is divided into $100 \times 100 = 100^2$ uniform grid points. Note that the Figure 5.12 includes the intervals that are

FIGURE 5.10
Distribution of the coverage probabilities of 95% confidence intervals for $n_1 = 20$ and $n_2 = 20$. KPM=Koopman, FM=Farrington and Manning MN=Miettinen and Nurminen WPF=Weighted profile likelihood, CZ=Chan and Zhang, AM=Agresti and Min.

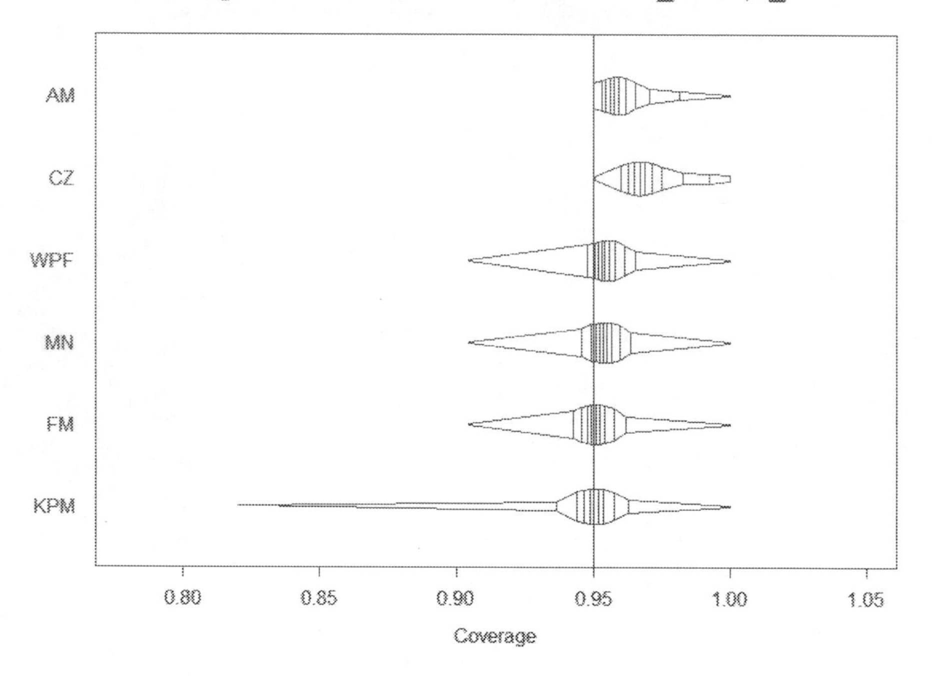

applicable for testing one-sided alternatives using TOST (see Chow, Shao and Wang [25]); thus the method such as **AM** is not included (see Santner et al. [80] for further details).

As expected, the **CZ** interval attains the nominal level for all grid points. This result confirms the findings of Santner et al. [80] that **CZ** interval always guarantees the nominal level, even for the ratio of proportions. Among the asymptotic intervals, **KPM** performed the worst. Most of its vertical bars are located to the left of the 95% level. SAS PROC FREQ'S default asymptotic option **AS** due to Gart and Nam [39] performs slightly better than the **KPM**. The asymptotic **MN** and the **WPF** (weighted profile likelihood) intervals perform reasonably well; most of its bars are near the 95% level, and it is less conservative (wider) than the **CZ** interval. Also, it appears that the **MN** and

FIGURE 5.11
Distribution of the expected lengths of 95% confidence intervals for $n_1 = 20$ and $n_2 = 20$. KZC=Katz et al. continuity corrected, KPM= Koopman, FM=Farrington and Manning MN=Miettinen and Nurminen WPF=Weighted profile likelihood, CZ=Chan and Zhang, AM=Agresti and Min.

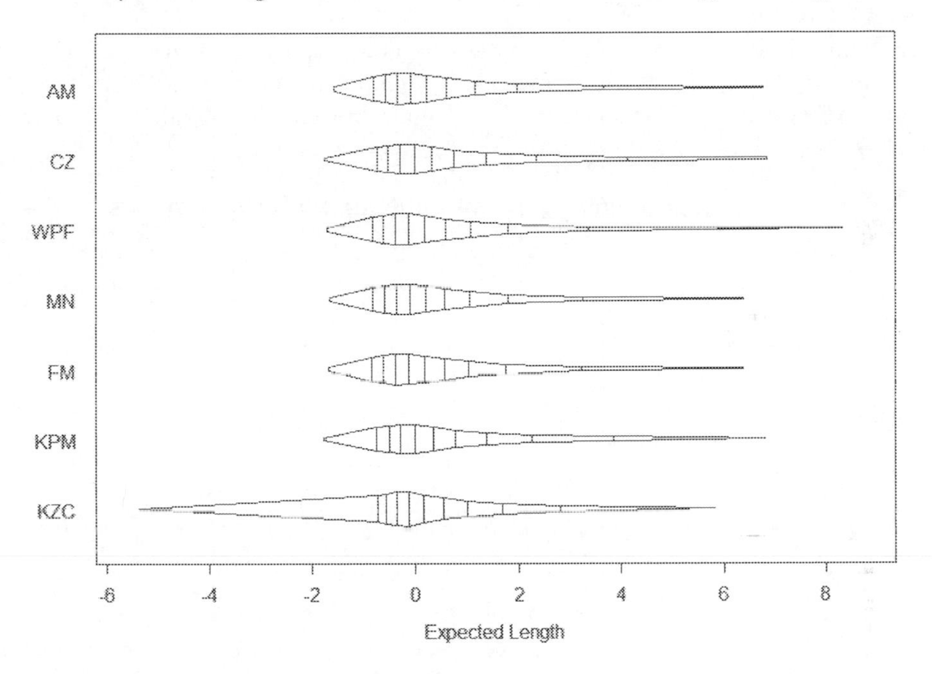

the **WPF** intervals perform well even for small sample sizes. However, further investigation is needed to confirm this finding.

FIGURE 5.12
Distribution of the coverage probabilities of 95% confidence intervals for $n_1 = 34$ and $n_2 = 34$. KPM=Koopman AS=asymptotic (Gart and Nam 1988) MN=Miettinen and Nurminen WPF=Weighted profile likelihood, CZ=Chan and Zhang.

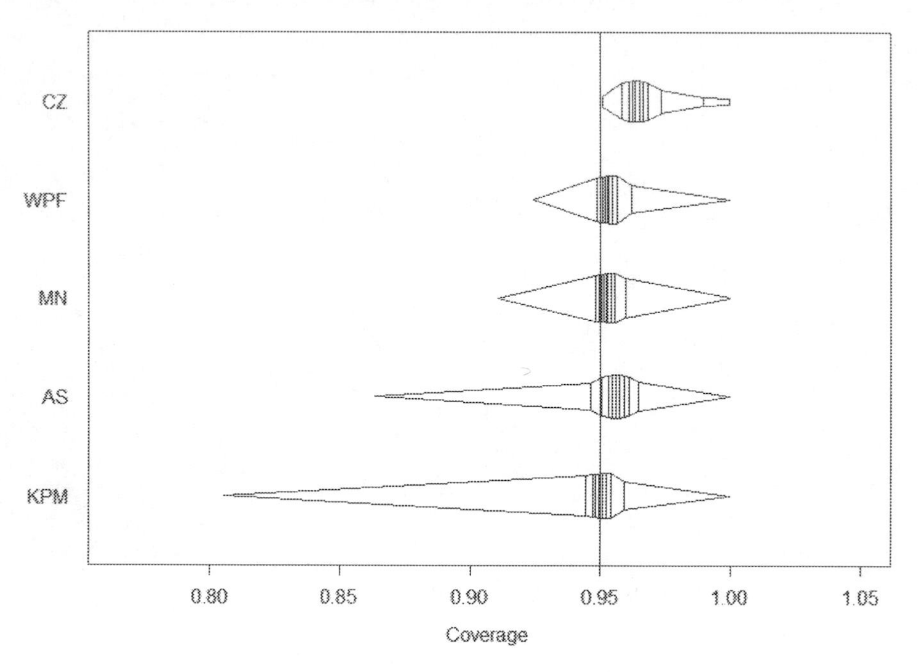

6

Paired Binomials: Difference of Proportions

6.1 Introduction

In biomedical research, more specifically in clinical trials, paired binary data arise frequently. Data obtained from matched case-control studies, crossover trials, and two arm observational studies with propensity score matching are examples of such data. A statistical analysis of paired binomial data must take into account the dependency introduced by matching.

To motivate the inference problem associated with paired binomial data, we start with an example. Consider the problem of testing equivalence of two diagnostic tests for malignancy, say, T_1 and T_2. Suppose Y_1 and Y_2 represent the results of the tests T_1 and T_2, respectively. For a patient, if $Y = 1$, the corresponding test result is positive (malignant), and $Y - 0$, if the test is negative. Given the test results (Y_1, Y_2) of n patients, the difference in proportions, say Δ, of positive results between two tests T_1 and T_2, is the parameter of interest. A confidence interval of Δ is often used in this context to investigate a statistical difference between the proportions.

There is considerable research on finding confidence intervals of the difference of proportions and carrying out relevant hypothesis testing procedures in a paired binomial setup. Newcombe [62] and Fagerland et al. [33] present excellent reviews of the available methods. In this chapter, we introduce the computational methods of the commonly used intervals of Δ based on paired data, compare their performances, and recommend their uses in practice.

6.2 Difference of two paired binomial proportions

Let Y_1 and Y_2 represent a pair of binary random variables, possibly correlated, similar to the scenario discussed above. Let $(y_{1i}, y_{2i}), i = 1, 2, ..., n$ be the test results for a random sample of n patients. Let us assume that there is no missing observation in the data. The data may be presented in the form of Table 6.1.

The joint distribution of (Y_1, Y_2) is given by $P(Y_1 = j, y_2 = k) = p_{jk}$, $(j, k) \in \{1, 0\}$. For a fixed $(j, k) \in \{1, 0\}$, let n_{jk} represent the number of

DOI: 10.1201/9781315169859-6

$Y_1 \rightarrow$	y_{11}	y_{12}	..	y_{1n}
$Y_2 \rightarrow$	y_{21}	y_{22}	..	y_{2n}

patients with $(y_{1i} = j, y_{2i} = k)$, and hence, $n_{11} + n_{10} + n_{01} + n_{00} = n$. The cell counts and the corresponding probabilities are summarized in Table 6.1.

TABLE 6.1
Paired 2 × 2 table

		Test B		
		1	**0**	
Test A	**1**	$n_{11}(p_{11})$	$n_{10}(p_{10})$	$n_{1+}(p_{1+})$
	0	$n_{01}(p_{01})$	$n_{00}(p_{00})$	$n_{0+}(p_{0+})$
		$n_{+1}(p_{+1})$	$n_{+0}(p_{+0})$	$n(1)$

Note that the proportions of positive result for Test A and Test B are $p_{1+} = p_{11} + p_{10}$ and $p_{+1} = p_{11} + p_{01}$, respectively. Also, n_{jk}/n, $(j,k) \in \{1,0\}$ represents the observed proportion of the test results in the cell (j,k). Let Δ be the difference of the probabilities of a positive result for Test A and Test B; therefore,

$$\Delta = (p_{11} + p_{10}) - (p_{11} + p_{01}) = p_{10} - p_{01}. \tag{6.1}$$

For n observed matched pairs, let $\boldsymbol{n} \equiv (n_{11}, n_{10}, n_{01}, n_{00})'$ represent the vector of cell counts. A natural estimate of the difference of proportions $(p_{10} - p_{01})$ is $(n_{10}/n - n_{01}/n)$, where n_{10} and n_{01} are the off-diagonal cell counts. Assuming that $\boldsymbol{n}$ has a multinomial distribution, the joint probability distribution of n_{10} and n_{01} is given by

$$\begin{aligned} f(n_{10}, n_{01}|n, \Delta, p_{01}) &= \frac{n!}{n_{10}!n_{01}!(n - n_{10} - n_{01})!}(p_{01} + \Delta)^{n_{10}} {p_{01}}^{n_{01}} \\ &\quad \times (1 - 2p_{01} - \Delta)^{n - n_{01} - n_{10}} \end{aligned} \tag{6.2}$$

where Δ and p_{01} satisfy the following conditions

$$\begin{cases} p_{01} \in [0, (1-\Delta)/2] & \text{if} \quad 0 < \Delta < 1, \\ p_{01} \in [-\Delta, (1-\Delta)/2] & \text{if} \quad -1 < \Delta < 0. \end{cases}$$

6.3 Hypotheses testing formulation

In the paired binomial setup, the hypotheses of interest are similar to the hypotheses discussed in Section 4.2.1. To bring an analogy with Section 4.2.1, let us denote p_{1+} and p_{+1} by p_1 and p_2, respectively. Hence $\Delta = p_1 - p_2$, and for a given $\Delta_0 > 0$, we wish to test either of the following three hypotheses:

1. The two-sided test:

$$H_0 : \Delta = 0 \quad \text{versus} \quad H_1 : \Delta \neq \Delta_0. \tag{6.3}$$

2. The one-sided upper-tailed test:

$$H_0 : \Delta \leq \Delta_0 \quad \text{versus} \quad H_1 : \Delta > \Delta_0. \tag{6.4}$$

3. The one-sided lower-tailed test:

$$H_0 : \Delta \geq \Delta_0 \quad \text{versus} \quad H_1 : \Delta < \Delta_0. \tag{6.5}$$

In clinical trial applications, one-sided tests are frequently used in testing for **non-inferiority** or **superiority** of a test drug compared to a standard drug. In testing for **non-inferiority**, the objective is to show that the test drug is not inferior to the standard drug by a certain pre-specified margin, say, Δ_0. Thus, the hypotheses for testing non-inferiority are:

$$H_0 : p_1 - p_2 \leq -\Delta_0 \quad \text{versus} \quad H_1 : p_1 - p_2 > -\Delta_0 \tag{6.6}$$

or

$$H_0 : \Delta \leq -\Delta_0 \quad \text{versus} \quad H_1 : \Delta > -\Delta_0 \tag{6.7}$$

where p_1 and p_2 are the probabilities of the cure for the test and the standard drug, respectively.

In (6.6), if $\Delta_0 < 0$, then the problem reduces to testing for superiority (see Chow, Shao and Wang [25]). Hence for $\Delta_0 > 0$, the hypotheses for testing superiority are:

$$H_0 : p_1 - p_2 \leq \Delta_0 \quad \text{versus} \quad H_1 : p_1 - p_2 > \Delta_0 \tag{6.8}$$

or

$$H_0 : \Delta \leq \Delta_0 \quad \text{versus} \quad H_1 : \Delta > \Delta_0 \tag{6.9}$$

In clinical trial applications, apart from testing for non-inferiority and superiority, testing for equivalence of two drugs is also very common. Testing for equivalence is used to demonstrate the similarities or lack of differences between two drugs against a pre-specified reference margin Δ_0. This is equivalent to testing simultaneously the two one-sided hypotheses given below(see Chow, Shao and Wang [25] for further details):

$$H_{0a} : p_1 - p_2 \leq -\Delta_0 \quad \text{versus} \quad H_{1a} : p_1 - p_2 > -\Delta_0, \tag{6.10}$$

$$H_{0b} : p_1 - p_2 \geq \Delta_0 \quad \text{versus} \quad H_{1b} : p_1 - p_2 < \Delta_0 \tag{6.11}$$

which in turn is equivalent to testing,

$$H_0 : |p_1 - p_2| \geq \Delta_0 \quad \text{versus} \quad H_1 : |p_1 - p_2| < \Delta_0 \tag{6.12}$$

Later in this chapter, we will illustrate testing of the above hypotheses using appropriate confidence intervals of Δ based on real-life data. In the following, we discuss different intervals of Δ for paired data along with the codes for computing such intervals.

6.4 Asymptotic intervals

Various asymptotic confidence intervals of Δ are proposed in the literature. These intervals are based on the asymptotic approximation of its coverage probability. In the following, we discuss the commonly used asymptotic intervals of Δ.

6.4.1 Wald interval

As stated above, in paired binomial setup the difference of proportions Δ is

$$\begin{aligned} \Delta &= p_{1+} - p_{+1} \\ &= (p_{11} + p_{10}) - (p_{11} + p_{01}) \\ &= p_{10} - p_{01} \end{aligned} \tag{6.13}$$

Hence a natural estimator of Δ is given by (cf. Table 6.1)

$$\begin{aligned} \widehat{\Delta} &= \widehat{p}_{10} - \widehat{p}_{01} \\ &= n_{10}/n - n_{01}/n, \end{aligned} \tag{6.14}$$

and an estimate of its variance is

$$\begin{aligned} \widehat{Var}(\widehat{\Delta}) &= \frac{(\widehat{p}_{10} + \widehat{p}_{01}) - (\widehat{p}_{01} - \widehat{p}_{10})^2}{n} \\ &= \frac{(n_{10} + n_{01}) - (n_{10} - n_{01})^2/n}{n^2}. \end{aligned}$$

Thus, the Wald interval with confidence coefficient $(1 - \alpha)$ is given by

$$\widehat{\Delta} \pm \frac{z_{\alpha/2}}{n}\sqrt{(n_{10} + n_{01}) - (n_{10} - n_{01})^2/n} \tag{6.15}$$

where $z_{\alpha/2}$ is the upper $100(\alpha/2)$ percentile point of the standard normal distribution. Note that when $n_{10} = n_{01} = 0$, the interval (6.15) reduces to the degenerate interval 0.

The R code for implementation of the Wald method is given below.

```
# Wald interval for binomial proportions with matched pairs
# "conflev"=confidence coefficient, n=sample size, b,c = off-diagonal counts
# b=n10 c=n01
diffpropci <- function(b,c,n,conflev)
{
  z  <- qnorm(1-(1-conflev)/2)
  diff <- (b-c)/n
  sd <- sqrt((b+c)-(c-b)^2/n)/n
  ll <- diff - z*sd
  ul <- diff + z*sd
  if(ll < -1) ll = -1
  if(ul > 1) ul = 1
  c(ll, ul)
}

#diffpropci(14,6,25,0.9)
```

The Wald interval is known to perform poorly in terms of coverage and expected lengths. As a remedy, the following continuity corrected Wald interval is proposed in the literature.

$$\widehat{\Delta} \pm \frac{z_{\alpha/2}}{n}\sqrt{(n_{10}+n_{01})-(n_{10}-n_{01})^2/n}+\frac{1}{n} \tag{6.16}$$

The R implementation of the continuity corrected Wald method is given in the following.

```
# Wald interval with continutity correction for binomial proportions with matched pairs
# "conflev"=confidence coefficient, n=sample size, b,c = off-diagonal counts
# b=n10 c=n01
diffpropci.cont <- function(b,c,n,conflev)
{
  z  <- qnorm(1-(1-conflev)/2)
  diff <- (b-c)/n
  sd <- sqrt((b+c)-(c-b)^2/n)/n +1/n
  ll <- diff - z*sd
  ul <- diff + z*sd
  if(ll < -1) ll = -1
  if(ul > 1) ul = 1
  c(ll, ul)
}

#diffpropci.cont(14,6,25,0.9)
```

The parameters p_{ij}, $i, j = 1, 0$ (cf. Table 6.1) can be equivalently expressed in terms of the parameters p_{1+}, p_{+1}, ϕ, where $\phi(= p_{11}p_{22}/p_{10}p_{01})$ is the odds ratio. Figure 6.1 shows the plots of the coverage probabilities of Wald interval and the continuity corrected Wald interval for $p_2(\equiv p_{+1})$, for varying values of $p_{1+}(\equiv p_1)$ when $p_2 = 0.5$, $n = 10$, and $\phi = 1$. Note that the Wald interval fails to attain the nominal coverage probability while the continuity corrected Wald interval attains it, but significantly above the nominal level (in fact, close to 100% level), thus leading to a very wide interval. In the ideal situation, one would expect the coverage probability to be fully aligned with the nominal level.

Figure 6.1 clearly demonstrates the limitations of the popular Wald confidence interval. The lack of reliability of Wald CI in terms of its coverage probability is disconcerting, to say the least, and makes the method unsuitable for critical applications in clinical research.

FIGURE 6.1
Coverage probabilities of Wald and continuity corrected Wald confidence intervals for $n = 10$ with $p_2 = 0.5$ and $\phi = 1$.

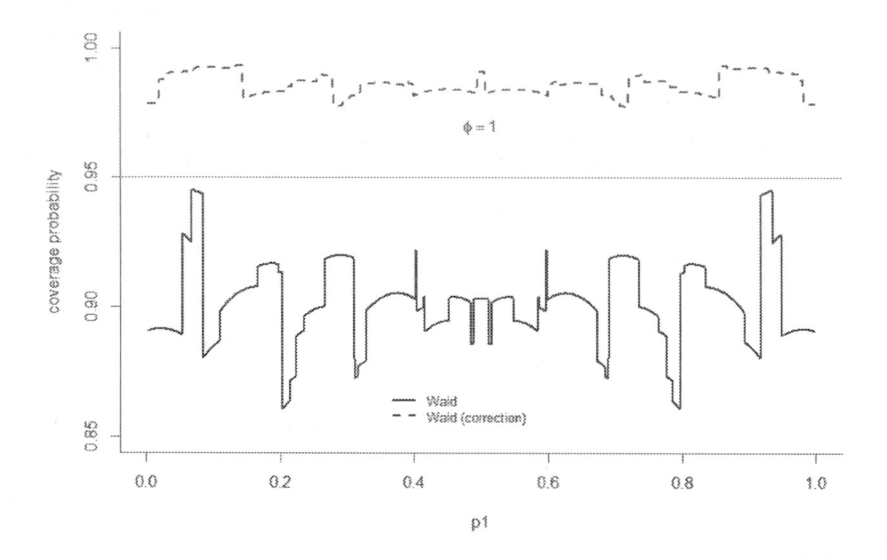

Illustration: Wald and continuity corrected Wald interval

The data considered here are from a study comparing two methods of detecting malignancy, namely, conventional biopsy and SpyGlass Direct Visualization system as discussed in Pradhan et al. [72]. The null hypothesis of interest is whether the probability of detecting malignancy using SpyGlass probe is less than or equal to that of the conventional biopsy. The data are given in Table 6.2.

Using the R code stated above, we run the following R script for computing the Wald and the continuity corrected Wald intervals at 90% level.

```
# Wald interval for the difference of binomial proportions with paired binomial data
```

TABLE 6.2
SpyGlass versus conventional biopsy

		Biopsy Results		
		Benign	Malignant	Total
SpyGlass Results	Benign	1	14	15
	Malignant	6	4	10
	Total	7	18	25

```
diffpropci(14,6,25,0.9)
# Wald interval with continuity correction
diffpropci.cont(14,6,25,0.9)
```

The R output:

```
> # Wald interval for the difference of binomial proportions with paired binomial data
> diffpropci(14,6,25,0.9)
[1] 0.04523558 0.59476442
> # Wald interval with continuity correction
> diffpropci.cont(14,6,25,0.9)
[1] -0.02055856  0.66055856
>
```

Note that the 90% continuity corrected Wald interval $(-0.02055856, 0.66055856)$ is much wider than the 90% Wald interval $(0.04523558, 0.59476442)$, confirming our observation from the simulation study reported above. The Wald interval excludes zero while the continuity corrected Wald interval includes it. Thus, the Wald interval leads to the rejection of the null hypothesis at 5% level of significance, while the continuity corrected Wald interval does not. As noted above, the attained coverage probability of the continuity corrected Wald interval consistently tends to be much higher than the nominal coverage probability. On the other hand, for the Wald interval, the coverage probability tends to be much lower than the nominal level. Thus, the conclusions drawn from these intervals may not be reliable and hence not recommended.

In the following, we discuss several intervals that improve the Wald interval's performance and are useful for clinical research.

6.4.2 Agresti and Min interval

Similar to the adjustment proposed for single binomial (Agresti and Coull [3]) and two independent binomials (Agresti and Caffo [1]), Agresti and Min [5] propose a simple adjustment to the Wald interval by adding 0.5 to each cell of the observed 2×2 table. Hence the $100(1-\alpha)\%$ Agresti and Min [5] interval is given by

$$\tilde{\Delta} \pm \frac{z_{\alpha/2}}{\tilde{n}} \sqrt{(\tilde{n}_{10} + \tilde{n}_{01}) - (\tilde{n}_{10} - \tilde{n}_{01})^2/\tilde{n}}, \tag{6.17}$$

where $\tilde{\Delta} = (\tilde{n}_{10} - \tilde{n}_{01})/\tilde{n}$, $\tilde{n}_{10} = n_{10} + 0.5$, $\tilde{n}_{01} = n_{01} + 0.5$, and $\tilde{n} = n + 2$. For this interval, the limits outside $[-1, 1]$ are truncated.

The R implementation of Agresti and Min interval is given in the following.

```
# Agresti and Min interval for difference of proportions with matched pairs
# "conflev"=confidence coefficient, n=sample size, b,c = off-diag counts
# b=n10 c=n01
diffpropci.AM <- function(b,c,n,conflev)
{
  z  <- qnorm(1-(1-conflev)/2)
  diff <- (b-c)/(n+2)
  sd <- sqrt((b+c+1)-(c-b)^2/(n+2))/(n+2)
  ll <- diff - z*sd
  ul <- diff + z*sd
  if(ll < -1) ll = -1
  if(ul > 1) ul = 1
  c(ll, ul)
}
```

Following the same set-up considered for Figure 6.1, we plot the attained coverage probabilities of the Agresti-min interval and the Wald interval in Figure 6.2. Note that the Agresti and Min interval is a much-improved version of the Wald interval, in terms of attaining the 95% nominal level.

FIGURE 6.2
Coverage probabilities of Wald and Agresti and Min confidence intervals for $n = 10$ with $p_2 = 0.5$ and $\phi = 1$.

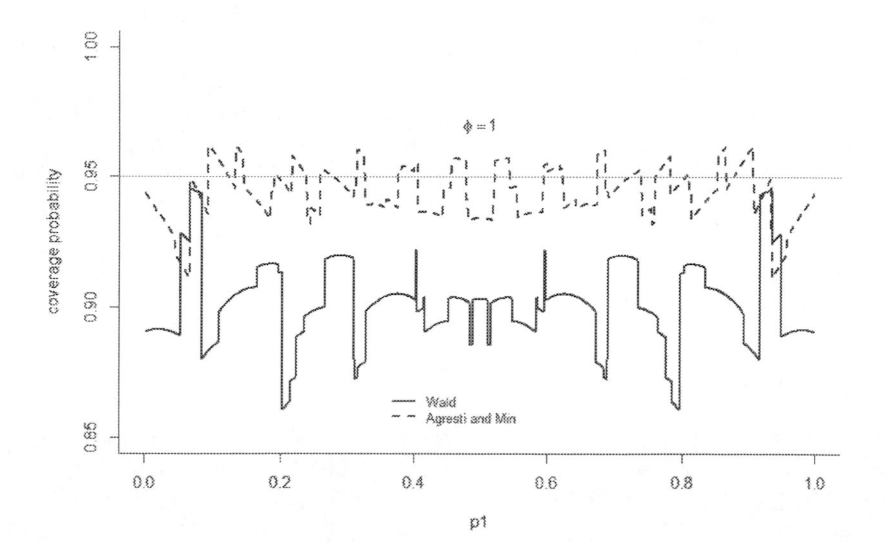

Illustration: Agresti and Min interval

Using the R code for computing the Wald interval, we run the following R script for computing the 90% Agresti-Min interval.

```
# # Agresti and Min for the difference of binomial proportions with paired
diffpropci.AM(14,6,25,0.9)
```

The R output is:

```
> # Agresti and Min for the difference of binomial proportions with paired
>diffpropci.AM(14,6,25,0.9)
[1] 0.03335088 0.55924171
>
```

Note that the 90% Agresti-Min interval is $(0.03335088, 0.55924171)$, and it excludes zero. Given the fact, even with smaller n ($n = 10$), the Agresti-Min interval attains the nominal coverage probabilities consistently for almost all values of p_1. Therefore, we can conclude that the null hypothesis is rejected at 5% level of significance.

6.4.3 MOVER intervals

Newcombe [62] proposes the "square-and-add" approach to finding the confidence interval of Δ. Later, Zou and Donner (2008) provided a theoretical justification for this approach in terms of local variance estimates recovered from the limits of the individual confidence intervals. The method is named MOVER (an acronym for the 'method of variance estimates recovery') by Zou (2008).

Suppose that the individual $100(1-\alpha)\%$ confidence intervals of p_{1+} and p_{+1} are (l_{1+}, u_{1+}) and (l_{+1}, u_{+1}), respectively, and $\hat{\rho}$ is an estimate of the correlation coefficient between $\hat{p}_{1+}$ and $\hat{p}_{+1}$, the estimates of p_{1+} and p_{+1}, respectively.

Then the 100(1-α)% MOVER interval for Δ is given by $(\underline{\Delta}, \overline{\Delta})$, where

$$\underline{\Delta} = \hat{p}_{1+} - \hat{p}_{+1} - \sqrt{(\hat{p}_{1+} - l_{1+})^2 + (u_{+1} - \hat{p}_{+1})^2 - 2\hat{\rho}(\hat{p}_{1+} - l_{1+})(u_{+1} - \hat{p}_{+1})}$$

and

$$\overline{\Delta} = \hat{p}_{1+} - \hat{p}_{+1} + \sqrt{(u_{1+} - \hat{p}_{1+})^2 + (\hat{p}_{+1} - l_{+1})^2 - 2\hat{\rho}(u_{1+} - \hat{p}_{1+})(\hat{p}_{+1} - l_{+1})}.$$

Tang et al. [89] discuss and compare different MOVER intervals for Δ obtained by plugging in l_{1+}, l_{+1}, u_{1+} and u_{+1} corresponding to different standard intervals for single binomial proportion in the above general formula and the following choices of $\hat{\rho}$.

A natural choice of $\hat{\rho}$ is an estimate of $\mathrm{Corr}(\hat{p}_{1+}, \hat{p}_{+1})$ which is given by

$$\hat{\rho}_1 = \frac{\hat{p}_{11}(1 - \hat{p}_{11}) - \hat{p}_{11}\hat{p}_{10} - \hat{p}_{11}\hat{p}_{01} - \hat{p}_{10}\hat{p}_{01}}{\sqrt{[\hat{p}_{11}(1 - \hat{p}_{11}) + \hat{p}_{10}(1 - \hat{p}_{10}) - 2\hat{p}_{11}\hat{p}_{10}][\hat{p}_{11}(1 - \hat{p}_{11}) + \hat{p}_{01}(1 - \hat{p}_{01}) - 2\hat{p}_{11}\hat{p}_{01}]}},$$

where, $\hat{p}_{ij}(=\frac{n_{ij}}{n})$'s is an estimate of p_{ij}. If n_{1+} (n_{+1}) equals to 0 or n, Tang et al. [89] suggest adding a value 0.5 to n_{10} (n_{01}) so that $\hat{\rho}_1$ is well-defined. The other choice of $\hat{\rho}$ is an estimate of standard Φ-coefficient in a 2×2 contingency table set-up given by (cf. Newcombe [62])

$$\hat{\rho}_2 = \frac{n_{11}n_{00} - n_{01}n_{10}}{\sqrt{n_{1+}n_{0+}n_{+1}n_{+0}}},$$

where $\hat{\rho}_2$ is equal to 0, if the denominator is 0, and the numerator is replaced by $\max(n_{11}n_{00} - n_{01}n_{10} - n/2, 0)$ if $n_{11}n_{00} - n_{01}n_{10} - n/2 > 0$. In the following, we specifically consider the MOVER intervals obtained from the Wilson (WI), the Agresti-Coull (AC), and the Jeffreys (JF) intervals.

6.4.3.1 MOVER Wilson interval

The 100 $(1-\alpha)$% MOVER-WI interval is obtained by plugging in the limits of the Wilson interval to the above general formula as shown below:

$$l_{1+} = \tilde{p}_{1+} - \frac{z_{\alpha/2}}{\bar{n}}\sqrt{n\hat{p}_{1+}(1-\hat{p}_{1+}) + \frac{z_{\alpha/2}^2}{4}}$$

$$u_{1+} = \tilde{p}_{1+} + \frac{z_{\alpha/2}}{\bar{n}}\sqrt{n\hat{p}_{1+}(1-\hat{p}_{1+}) + \frac{z_{\alpha/2}^2}{4}}$$

and

$$l_{+1} = \tilde{p}_{+1} - \frac{z_{\alpha/2}}{\bar{n}}\sqrt{n\hat{p}_{+1}(1-\hat{p}_{+1}) + \frac{z_{\alpha/2}^2}{4}}$$

$$u_{+1} = \tilde{p}_{+1} + \frac{z_{\alpha/2}}{\bar{n}}\sqrt{n\hat{p}_{+1}(1-\hat{p}_{+1}) + \frac{z_{\alpha/2}^2}{4}}$$

where,

$$\tilde{p}_{1+} = \frac{n_{1+} + 0.5z_{\alpha/2}^2}{n + z_{\alpha/2}^2}, \quad \tilde{p}_{+1} = \frac{n_{+1} + 0.5z_{\alpha/2}^2}{n + z_{\alpha/2}^2}; \quad \bar{n} = n + z_{\alpha/2}^2$$ and $z_{\alpha/2}$ is the

$$100(\alpha/2)$$

upper percentile of the standard normal distribution.

6.4.3.2 MOVER Agresti–Coull interval

The 100(1-α)% MOVER-AC interval is obtained by plugging in

$$l_{1+} = \tilde{p}_{1+} - z_{\alpha/2}\sqrt{\frac{\tilde{p}_{1+}(1-\tilde{p}_{1+})}{\bar{n}}}; \quad u_{1+} = \tilde{p}_{1+} + z_{\alpha/2}\sqrt{\frac{\tilde{p}_{1+}(1-\tilde{p}_{1+})}{\bar{n}}}$$

and

$$l_{+1} = \tilde{p}_{+1} - z_{\alpha/2}\sqrt{\frac{\tilde{p}_{+1}(1-\tilde{p}_{+1})}{\bar{n}}}; \quad u_{+1} = \tilde{p}_{+1} - z_{\alpha/2}\sqrt{\frac{\tilde{p}_{+1}(1-\tilde{p}_{+1})}{\bar{n}}}$$

to the general formula.

6.4.3.3 MOVER Jeffreys' interval

The 100(1-α)% MOVER-JF interval is obtained by plugging in

$$l_{1+} = \frac{2n_{1+}+1}{2n_{1+}+1+[2(n-n_{1+})+1]F_{\alpha/2}(2(n-n_{1+})+1, 2n_{1+}+1)},$$

$$u_{1+} = \frac{2n_{1+}+1}{2n_{1+}+1+[2(n-n_{1+})+1]F_{1-\alpha/2}(2(n-n_{1+})+1, 2n_{1+}+1)}$$

and

$$l_{+1} = \frac{2n_{+1}+1}{2n_{+1}+1+[2(n-n_{+1})+1]F_{\alpha/2}(2(n-n_{+1})+1, 2n_{+1}+1)},$$

$$u_{+1} = \frac{2n_{+1}+1}{2n_{+1}+1+[2(n-n_{+1})+1]F_{1-\alpha/2}(2(n-n_{+1})+1, 2n_{+1}+1)},$$

to the general formula, where, $F_\gamma(\nu_1, \nu_2)$ is the upper 100γ-th percentile of the F-distribution with ν_1 and ν_2 degrees of freedom. For further details, see Tang et al. [89].

The following is the SAS/IML implementation of the MOVER method.

```
/*-------implementation of Tang et al. (2010)-----------------------------*/
/*Specifying type=1 produces Tang et al CI adjusting marginal theta_hats  */
/*Specifying type=0 produces Newcombe CI un-adjusting marginal theta hats */
%macro comBin(type=,a= ,b= , c= , n= , alpha= );

/*Creating binomila CIs for theta1*/
data one;
x=1; y=1; w=%eval(&a+&b);output;
x=1; y=2; w=%eval(&n-(&a+&b));output;
;

run;
ods noresults;
ods listing close;
ods output  Freq.Table1.BinomialCLs=CIs;
proc freq data=one;
tables y/binomial (all) alpha=&alpha;
weight w/zeros;
run;

data CI1; set CIs;
n=_n_;
if Type="Clopper-Pearson (Exact)"|Type="Wald" then delete;
drop n;
```

```
run;


/*p2 computation*/
data one;
x=1; y=1; w=%eval(&a+&c);output;
x=1; y=2; w=%eval(&n-(&a+&c));output;
;

run;
/*Creating binomila CIs for theta2*/
ods output  Freq.Table1.BinomialCLs=CI2s;
proc freq;
tables y/binomial (all) alpha=&alpha;
weight w/zeros;
run;


data CI2; set CI2s;
n=_n_;
if Type="Clopper-Pearson (Exact)"|Type="Wald" then delete;
drop n;
run;


proc iml;
use CI1;
read all var{LowerCL} into l1;
read all var{UpperCL} into u1;
use CI2;
read all var{LowerCL} into l2;
read all var{UpperCL} into u2;


n=%eval(&n);a= %eval(&a); b=%eval(&b); c=%eval(&c); d=n-(a+b+c);

%if &type=1 %then %do;
p11=a/n;p10=b/n;p01=c/n;
if(b=0 & c=0) then do;
p10=0.5/n;
p01=0.5/n;
end;

D1=(p11*(1-p11)+p10*(1-p10)-2*p11*p10)*(p11*(1-p11)+p01*(1-p01)-2*p11*p01);

if D1<0 then D1=1e-6;

N1=p11*(1-p11)-p11*p10-p11*p01-p10*p01;

corr2=N1/sqrt(D1);

/*Marginal theta_hat corrections based on AM method*/
theta3 = (a+b+0.5*(probit(1-0.05/2)##2))/(n+probit(1-0.05/2)##2);
theta4 = (a+c+0.5*(probit(1-0.05/2)##2))/(n+probit(1-0.05/2)##2);

%end;

%else %do;

theta3 = (a+b)/(n);
theta4 = (a+c)/(n);
denom=sqrt((a + b)* (c + d) * (a + c) * (b + d));

if denom=0 then corr2=0;
else if (a*d > b*c)then do;
corr2= max(((a * d - b * c - n/2) / sqrt((a + b) * (c + d) * (a + c) * (b + d))), 0);
end;
```

```
else do;
corr2 = ((a * d - b * c)/denom );
end;

%end;

theta1=repeat(theta3,3,1);
theta2=repeat(theta4,3,1);
print theta1 theta2;
corr=corr2;

A1=j(3,1,-1) ;
/*Final checking for exceeding limits -1 and 1 - j(3,1,-1) or j(3,1,1) used;*/
L =A1<>((theta1 - theta2) - sqrt((theta1 - l1)##2 + (u2 - theta2)##2
- 2 #corr # (theta1 - l1)# (u2 - theta2)));

U =j(3,1,1) ><((theta1 - theta2) + sqrt((u1 - theta1)##2 + (theta2 - l2)##2
- 2# corr# (u1 - theta1)# (theta2 - l2)));

estimates=L||U;
create correct from estimates [colname=('Lower'||'Upper') ] ;
append from estimates;
close correct;

quit;
data typ; set Ci1;
Keep Type;
run;

data combo;
merge typ correct;
run;
ods listing;
options nodate nonumber;
proc print data=combo noobs ;
%if &type=1 %then %do;
title "MOVER CI of the difference of two paired binomial proportions -- Tang et al.";
%end;
%else %do;
title "MOVER CI of the difference of two paired binomial proportions -- Newcombe";
%end;
run;
%mend;
```

Figure 6.3 shows the plots of the coverage probabilities of Wilson, Agresti and Coull and Jeffreys intervals for $p_2 (\equiv p_{+1})$, for varying values of $p_{1+} (\equiv p_1)$ when $p_2 = 0.5$, $n = 10$, and $\phi = 1$.

Note that all three intervals have coverage probabilities closely aligned with the 95% nominal levels for varying values of p_1 except near $p_1 = 0$ and $p_1 = 1$.

Illustration: MOVER intervals

Section 6.3 discusses different statistical hypothesis testing problems that often arise in real-life applications. For example, in crossover trials, testing the equivalence of two drugs in terms of their effects is often of interest. For illustration, we consider a data set given in Tango [91], where 44 participants

FIGURE 6.3
Coverage probabilities of MOVER confidence intervals for $n = 10$ with $p_2 = 0.5$ and $\phi = 1$.

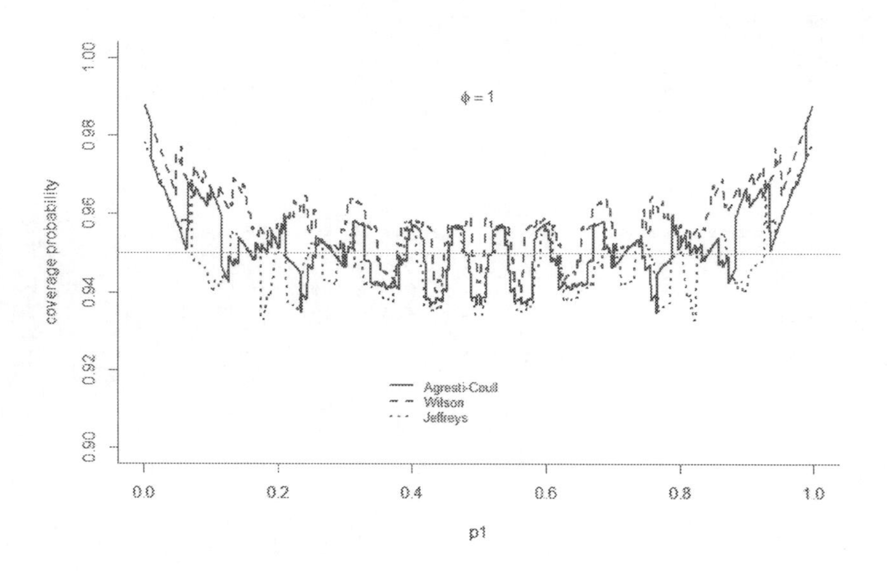

were randomly assigned to two treatment sequences AB or BA; A represents the assessment of soft contact lenses using chemical disinfection, and B using the thermal disinfection system. The data are shown in Table 6.3.

TABLE 6.3
Cross-over trial data of disinfection treatments for soft contact lenses

		Thermal disinfection		
		Effective	**Ineffective**	**Total**
Hydrogen peroxide	**Effective**	43	0	43
	Ineffective	1	0	1
	Total	44	0	44

The null hypothesis to be tested is that the two treatments are not equivalent within a margin of $\Delta_0 = 0.1$ with one-sided 2.5% significance level. Hence equation 6.12entails that the null hypothesis is to be rejected at 5% level of significance if the 95% confidence interval of Δ lies within the interval $[-0.1, 0.1]$. We now run the following SAS code using $\hat{\rho}_2$.

```
/*-------implementation of Tang et al. (2010)----------------------------*/
/*Specifying type=1 produces Tang et al CI adjusting marginal theta_hats  */
/*Specifying type=0 produces Newcombe CI un-adjusting marginal theta_hats */

%comBin(type=0,a=43 ,b=0 , c=1 , n=44 , alpha=0.05 );
```

The SAS output is:

```
MOVER CI of the difference of two paired binomial proportions -- Newcombe

                    Type                Lower         Upper

                    Agresti-Coull     -0.12889     0.075706
                    Jeffreys          -0.10134     0.036059
                    Wilson            -0.11808     0.059718
```

Note that the lower limits of all the intervals above are less than -0.1. Thus, the MOVER intervals considered above fail to reject the null hypothesis at 5% level of significance confirming that the two disinfecting treatments are not equivalent.

Now, we test the same hypothesis by running the SAS code for the MOVER intervals using $\hat{\rho}_1$.

```
/*-------implementation of Tang et al. (2010)-------------------*/
/*Specifying type=1 produces Tang et al CI adjusting marginal theta_hats  */
/*Specifying type=0 produces Newcombe CI un-adjusting marginal theta_hats */

%comBin(type=1,a=13 ,b=0 , c=1 , n=44 , alpha=0.05 );
```

The SAS shows the following output:

```
MOVER CI of the difference of two paired binomial proportions -- Tang et al

                   Type                Lower          Upper

                   Agresti-Coull     -0.099735     0.061689
                   Jeffreys          -0.077778     0.039583
                   Wilson            -0.090644     0.048839
```

Note that all the MOVER intervals lie within $[-0.1, 0.1]$. Therefore, the null hypothesis is rejected at 5% level of significance, contradicting the conclusion arrived at by using $\hat{\rho}_2$ above. Clearly, in terms of statistical significance, both sets of intervals should lead to a very similar conclusion if one does not stick blindly to the 5% significance rule.

6.4.4 Asymptotic score interval

For the paired binomial setup, Tango [91] propose an asymptotic score interval. For testing the null hypothesis $\Delta = p_{1+} - p_{+1} = \Delta_0$, the score statistic is

$$T(\boldsymbol{n}|\Delta_0) = \frac{n_{10} - n_{01} - n\Delta_0}{\sqrt{n[2\tilde{p}_{01} + \Delta_0(1 - \Delta_0)]}} \tag{6.18}$$

where, $\tilde{p}_{01}$, is the maximum likelihood estimate of p_{01} under the constraint $p_{1+} - p_{+1} = \Delta_0$. It is given by

$$\tilde{p}_{01} = \frac{-b + \sqrt{b^2 - 4ac}}{2a} \tag{6.19}$$

where $a = 2n$, $b = (2n + n_{10} - n_{01})\Delta_0 - n_{10} - n_{01}$ and $c = -n_{01}\Delta_0(1 - \Delta_0)$. Thus, the $100(1-\alpha)\%$ confidence interval $(\underline{\Delta}, \overline{\Delta})$ is obtained by solving the following equations iteratively,

$$T(\boldsymbol{n}|\underline{\Delta}) = z_{\alpha/2},$$

and

$$T(\boldsymbol{n}|\overline{\Delta}) = -z_{\alpha/2}.$$

The R implementation of this method is available on Agresti's website (`http://www.stat.ufl.edu/~aa/cda/R/`). One can also use the R-package *ratesci* from *C*RAN to compute this interval. The non-iterative version of the interval is also available in (`https://works.bepress.com/zyang/1/`).

```
#---------Tango (score) interval for the difference of paired binomial proportions
#---------n10, n01 are the cell counts of the off-diagonal elements, n is the total
# --------conflev is the confidence level----------------------------------------
TGscoreci <- function(n10,n01,n,conflev)
{
  pa = 2*n
  z = qnorm(1-(1-conflev)/2)

  if(n01 == n) {ul = 1}
  else{
    proot = (n01-n10)/n
    dp = 1-proot
    niter = 1
    while(niter <= 50){
      dp = 0.5*dp
      up2 = proot+dp
      pb = - n10 - n01 + (2*n-n01+n10)*up2
      pc = -n10*up2*(1-up2)
      q21 = (sqrt(pb^2-4*pa*pc)-pb)/(2*pa)
      score = (n01-n10-n*up2)/sqrt(n*(2*q21+up2*(1-up2)))
      if(abs(score)<z){ proot = up2 }
      niter=niter+1
      if((dp<0.0000001) || (abs(z-score)<.000001)){
        niter=51
        ul=up2
      }
    }
  }

  if(n10 == n) {ll = -1}
  else{
    proot = (n01-n10)/n
    dp = 1+proot
    niter = 1
    while(niter <= 50){
      dp = 0.5*dp
      low2 = proot-dp
      pb = - n10 - n01 + (2*n-n01+n10)*low2
```

```
        pc = -n10*low2*(1-low2)
        q21 = (sqrt(pb^2-4*pa*pc)-pb)/(2*pa)
        score = (n01-n10-n*low2)/sqrt(n*(2*q21+low2*(1-low2)))
        if(abs(score) < z){proot = low2}
        niter = niter+1
        if((dp<0.0000001) || (abs(z-score)<.000001)){
          ll = low2
          niter = 51
        }
      }
    }
    c(ll,ul)
}
```

Under the same set-up considered for Figure 6.1, we plot the attained coverage probabilities of the Tango interval in Figure 6.4. Note that the asymptotic score interval of Tango seems to have very good coverage properties for almost all the values of p_1.

FIGURE 6.4
Coverage probabilities of Tango (score) confidence intervals for $n = 10$ with $p_2 = 0.5$ and $\phi = 1$.

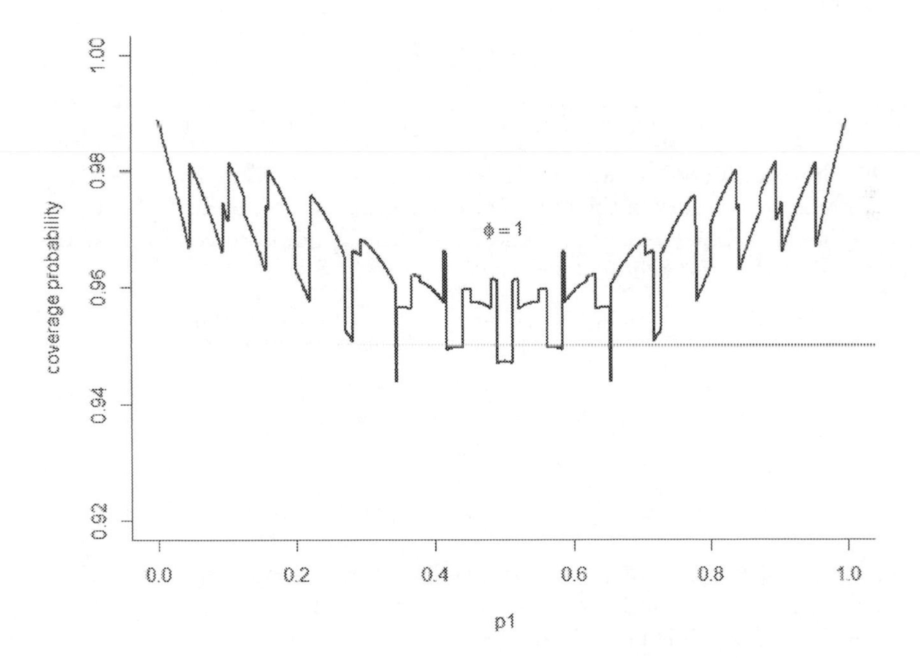

Illustration: Tango interval

The Tango interval applied to the data set in Table 6.3 is given below.

```
#---------Tango (score) interval for the difference of paired binomial proportions
#---n10, n01 are the cell counts of the off-diagonal elements, n is the total
# --------conflev is the confidence level--------------------------------------

TGscoreci(1,0,44,0.95)
```

Following is output in R console:

```
> TGscoreci(1,0,44,0.95)
[1] -0.11807715  0.05939325
>
```

Since the lower limit is less than -0.1, the null hypothesis cannot be rejected, and plausibly, the two disinfection treatments are not equivalent.

6.4.5 Weighted profile likelihood interval

Consider the set-up as in Table 6.1. Defining $\Delta = p_{1+} - p_{+1} = p_{10} - p_{01}$ the joint probability distribution of (n_{10}, n_{01}) can be written as

$$\begin{aligned} f(n_{10}, n_{01}|\Delta, p_{01}) &= \frac{n!}{n_{10}!n_{01}!(n - n_{10} - n_{01})!} \\ &\quad \times (p_{01} + \Delta)^{n_{10}} p_{01}^{n_{01}} (1 - 2p_{01} - \Delta)^{n - n_{10} - n_{01}}, \end{aligned}$$

where, $0 \leq p_{01} \leq (1 - \Delta)/2$, if $0 < \Delta < 1$ and $-\Delta \leq p_{01} \leq (1 - \Delta)/2$, if $-1 < \Delta < 0$. Thus, one can construct a likelihood-based interval for Δ based on the above likelihood function. However, Agresti and Min [5] suggest that the coverage probability of the interval can be improved by adding $a/4$ $(a > 0)$ to each of the cell frequencies of Table 6.1. With the adjusted cell frequencies, the kernel of the log-likelihood becomes

$$\begin{aligned} l(\Delta, p_{01}) &\propto (n_{10} + a/4)\ln(p_{01} + \Delta) + (n_{01} + a/4)\ln p_{01} \\ &\quad + (n - n_{10} - n_{01} + a/2)\ln(1 - 2p_{01} - \Delta). \end{aligned}$$

Thus, the kernel of the weighted profile log-likelihood becomes

$$\begin{aligned} l(\Delta, \tilde{p}_{01}) &\propto (n_{10} + a/4)\ln(\tilde{p}_{01} + \Delta) + (n_{01} + a/4)\ln \tilde{p}_{01} \\ &\quad + (n - n_{10} - n_{01} + a/2)\ln(1 - 2\tilde{p}_{01} - \Delta), \end{aligned}$$

where $\tilde{p}_{01}$ is the restricted maximum likelihood estimate of p_{01} for given Δ. Therefore, the approximate 100(1 - α)% weighted profile likelihood-based confidence interval for Δ is given by

$$\{\Delta : 2[l(\hat{\Delta}, \hat{p}_{01}) - l(\Delta, \tilde{p}_{01})] \leq \chi_1^2(\alpha)\},$$

where $\hat{\Delta}$ and $\hat{p}_{01}$ are the unrestricted maximum likelihood estimates of Δ and p_{01}, respectively, and $\chi_1^2(\alpha)$ is the upper 100α percentile point of the chi-squared distribution with 1 degree of freedom. Following Venzon and Moolgavkar [94], the confidence interval for Δ, say, (Δ_L, Δ_U), is the admissible solution to the following system of non-linear equations:

$$\begin{bmatrix} l(\hat{\Delta}, \hat{p}_{01}) - l(\Delta, p_{01}) - \frac{1}{2}\chi_1^2(\alpha) \\ \frac{\partial l(\Delta, p_{01})}{\partial p_{01}} \end{bmatrix} = 0.$$

Agresti and Min [5] point out that there is no justification for a specific choice of a because of its ad hoc nature. However, using intensive simulation studies, Pradhan et al. [72] find that $a = 2$ performs well in most of the practical situations.

The following is the SAS/IML implementation of the above method.

```
/*where a=n11 b=n10 c=n01 n=n11+n0+n01+n00 */
%macro Pf(n= ,b= ,c=,a= , alpha= );
proc iml;
/*getting the loglikelihood*/
Start likelihood(x)global(b,c,n) ;
L=b#log(x[1]+x[2])+c#log(x[1])+(n-b-c)#log(1-2#x[1]-x[2]);
return (L);
finish likelihood;
/*getting the gradient vectors*/
start Gradient(x)global(b,c,n);
g=j(1,2,1e-16);
g[1]=b/(x[1]+x[2])+c/x[1]+(n-b-c)#(-2)/(1-2#x[1]-x[2]);
g[2]=b/(x[1]+x[2])+(n-b-c)#(-1)/(1-2#x[1]-x[2]);
return (g);
finish Gradient;
/*adjusting the cell frequencies*/
 b=%sysevalf(&b+(&a/4)) ;c=%sysevalf(&c+(&a/4)); n=%sysevalf(&n+&a);
 x0 ={1e-6 0.9999};
 optn = {1 0};
 con = { 1.e-6 -0.99999 . .,
 0.99999 0.99999 . . ,
1 1 1 1e-6,
1 1 -1 0.99999};
/*Calling SAS's optimization function*/
 call nlptr(rc,xres,"likelihood",x0,optn,con,,,,"Gradient");
 xopt = xres`; fopt = likelihood(xopt);

 /*Calling SAS's optimization function to get hessian*/
 call nlpfdd(f,g,hes2,"likelihood",xopt,,"Gradient");

 start plgrad(x) global(like,ipar,lstar);
 like = likelihood(x);
 grad = Gradient(x);
 grad[ipar]=like-lstar;
 return(grad`);
 finish plgrad;

prob=&alpha ;
  xlb=j(2,1,-1);
xub=j(2,1,1);
/* quantile of chi**2 distribution */
chqua = cinv(1-prob,1);like=fopt; lstar = fopt - .5 * chqua;

 optn = {2 0};
 do ipar = 1 to 2;
 /* Implementing of Venzon & Moolgavkar (1988)*/
 if ipar=1 then ind = 2; else ind = 1;
```

```
 delt = - inv(hes2[ind,ind]) * hes2[ind,ipar];
 alfa = - (hes2[ipar,ipar] - delt' * hes2[ind,ipar]);
 if alfa > 0 then alfa = .5 * sqrt(chqua / alfa);
 else do;
 print "Bad alpha";
 alfa = .1 * xopt[ipar];
 end;
 if ipar=1 then delt = 1 || delt;
 else delt = delt || 1;
 x0 = xopt + (alfa * delt)';
 con2 = con; con2[1,ipar] = xopt[ipar];
 tc={2000,5000};
 call nlplm(rc,xres,"plgrad",x0,optn,con2,tc );
 f = plgrad(xres); s = ssq(f);
 if (s <1.e-6) then xub[ipar] = xres[ipar];
  else xub[ipar] =1.0;
 x0 = xopt - (alfa * delt)';
 con2[1,ipar] = con[1,ipar]; con2[2,ipar] = xopt[ipar];
 tc={2000,5000};
 call nlplm(rc,xres,"plgrad",x0,optn,con2,tc);
 f = plgrad(xres); s = ssq(f);
 if (s < 1.e-6) then xlb[ipar] = xres[ipar];
 else xlb[ipar] = -1.0;
 end;

ci=xlb||xub;
create Profile from ci[colname={'LowerPF','UpperPF'} ] ;
append from ci;
close Profile;
run;
quit;
run;

data Profile; set Profile;
if _n_=1 then delete;
LowerCI=round(LowerPF, 0.0001);
UpperCI=round(UpperPF, 0.0001);
drop LowerPF UpperPF;
run;
PROC Print data=Profile noobs;
title "Confidence interval of the difference of two paired binomials\\
based on penalized profile likelihood";
run;
%mend;


/*Results for AS05*/
options nodate nonumber;
%Pf(n=285 ,b=11 ,c=3,a=2, alpha=0.05 );
/*Results for AS1*/
%Pf(n=285 ,b=11 ,c=3,a=4,alpha=0.05 );
```

We plot the attained coverage probabilities of the weighted profile likelihood interval in Figure 6.5 for the adjustments $0.5(a = 2)$ and $1(a = 4)$ in the same setup as considered for Figure 6.1. Note that the adjustment 0.5 ($a = 2$ in the SAS macro) to each cell of the observed data yields a much-improved interval in terms of attained coverage probability except for the values of p_1 near 0 and 1.

Illustration: Weighted profile likelihood interval

Consider the SpyGlass study example (Table 6.2) stated in section 6.4.1. The null hypothesis of interest is whether the probability of detecting malignancy

FIGURE 6.5
Coverage probabilities of weighted profile likelihood confidence intervals for $n = 10$ with $p_2 = 0.5$ and $\phi = 1$.

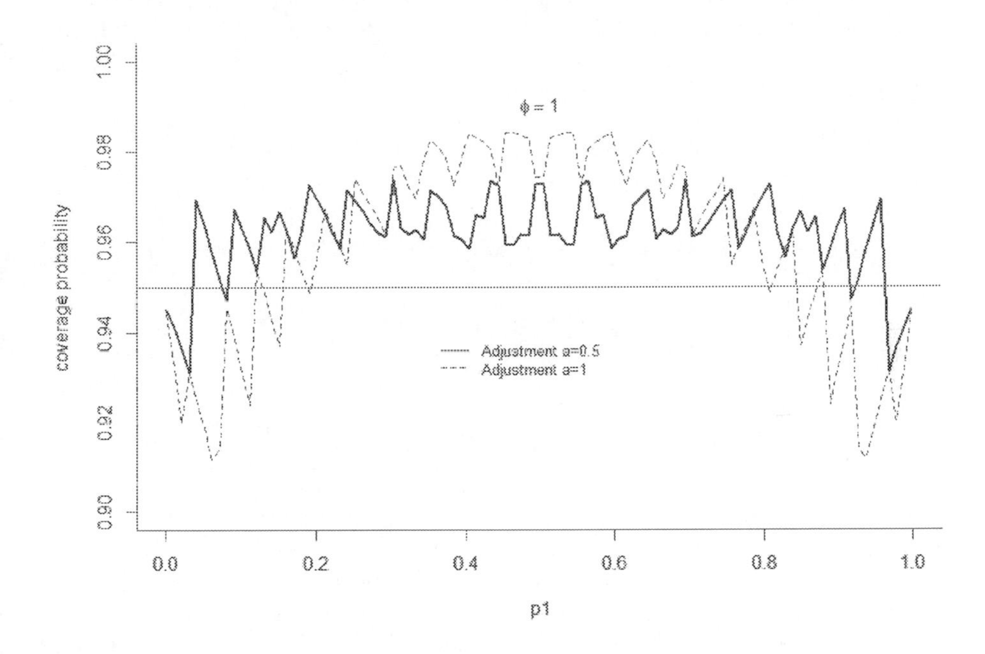

using SpyGlass probe is less than equal to that of the conventional biopsy. To test the hypothesis, run the following SAS code for 90% CI using the SAS/Macro included before.

```
ods listing;
options nodate nonumber;
%Pf(n=25 ,b=14 ,c=6,a=2, alpha=0.1 );
```

The SAS output:

```
Confidence interval of the difference of two paired binomials\\based on pen

                                Lower    Upper
                                 CI        CI

                                0.0212    0.539
```

Note that the 90% Weighted profile likelihood interval excludes zero, hence the null hypothesis is rejected confirming that the probability of detecting malignancy using SpyGlass probe is not less than the conventional biopsy method.

6.4.6 Confidence interval based on bivariate copula

To define a bivariate binomial distribution of n_{1+} and n_{+1} (cf. Table 6.1), we proceed as follows. Let (Y_{1i}, Y_{2i}), $i = 1, 2, ..., n$ be n independent pairs of binary variables each following a bivariate Bernoulli distribution with a dependence parameter, say α. Marginally y_{1i} and y_{2i} have Bernoulli distributions with parameters π_1 and π_2, respectively. The objective is to find a confidence interval of $\Delta = \pi_1 - \pi_2$.
We introduce the following notations:

$$\theta = (\pi_1, \pi_2, \alpha); P(Y_1 = y_1, Y_2 = y_2) = f(y_1, y_2; \theta)$$

$$P(Y_1 \leq y_1, Y_2 \leq y_2) = F_Y(y_1, y_2; \theta)$$

$$F_Y(y_1, y_2; \theta) = C_\alpha(F_{Y_1}(y_1), F_{Y_2}(y_2))$$

where

$$\begin{array}{l} F_{Y_1}(y_1) = P\{Y_1 \leq y_1\} = \sum_{r=0}^{y_1} \binom{n}{r} \pi_1^r (1 - \pi_1)^{(n-r)} \\ F_{Y_2}(y_2) = P\{Y_2 \leq y_2\} = \sum_{r=0}^{y_2} \binom{n}{r} \pi_2^r (1 - \pi_2)^{(n-r)} \end{array}$$

and $C_\alpha(u_1, u_2)$ is a copula function with dependence parameter α. One can choose different types of copula functions to get the estimates of α. In the paired binomial setup, the dependence can be positive as well as negative. Therefore, it is natural to consider a copula whose dependence parameter supports negative as well as a positive dependence in the entire parameter space. As a natural choice, similar to Meester and MacKay (1994), we consider the Frank copula, given by,

$$C_\alpha(u_1, u_2) = -\alpha^{-1} log\{1 + \frac{(exp(-\alpha u_1) - 1)(exp(-\alpha u_2) - 1)}{(exp(-\alpha) - 1)}, \quad (6.20)$$

where $0 < u_1, u_2 < 1, -\infty < \alpha < \infty$. The parameter $\alpha \to -\infty$ implies extreme negative dependence, $\alpha \to \infty$ implies extreme positive dependence, and $\alpha = 0$ implies independence. The Frank copula captures weak tail dependence but strong dependence in the middle of the distribution.

Thus, for the paired binary data (Y_{1i}, Y_{2i}), $i = 1, 2, .., n$, using the distribution defined above, one can write the following:

$$\begin{aligned} p_{00} &= P(Y_1 = 0, Y_2 = 0) \\ &= F_Y(0, 0; \theta) \\ &= C_\alpha(1 - \pi_1, 1 - \pi_2), \end{aligned}$$

$$\begin{aligned} p_{10} &= P(Y_1 = 1, Y_2 = 0) \\ &= F_Y(1, 0; \theta) - F_Y(0, 0; \theta) \\ &= C_\alpha(1, 1 - \pi_2) - C_\alpha(1 - \pi_1, 1 - \pi_2), \end{aligned}$$

$$\begin{aligned} p_{01} &= P(Y_1 = 0, Y_2 = 1) \\ &= F_Y(0, 1; \theta) - F_Y(0, 0; \theta) \\ &= C_\alpha(1 - \pi_1, 1) - C_\alpha(1 - \pi_1, 1 - \pi_2), \end{aligned}$$

and

$$\begin{aligned} p_{11} &= P(Y_1 = 1, Y_2 = 1) \\ &= F_Y(1, 1; \theta) - F_Y(0, 1; \theta) - F_Y(1, 0; \theta) + F_Y(0, 0; \theta) \\ &= C_\alpha(1, 1) - C_\alpha(1 - \pi_1, 1) - C_\alpha(1, 1 - \pi_2) + C_\alpha(1 - \pi_1, 1 - \pi_2). \end{aligned}$$

Therefore, for Table 6.1, the likelihood function can be written as

$$L(x; \theta) \propto p_{11}^{n_{11}} p_{10}^{n_{10}} p_{01}^{n_{01}} p_{00}^{n_{00}}. \tag{6.21}$$

As above, incorporating the Agresti and Min [5] adjustment of cell frequencies the kernel of the weighted log-likelihood function can be written as

$$\begin{aligned} l(x; \pi_1, \pi_2) \propto\ & (n_{11} + a/4)\ln(p_{11}) + (n_{10} + a/4)\ln(p_{10}) + (n_{01} \\ & + a/4)\ln(p_{01}) + (n_{00} + a/4)\ln(p_{00}) \end{aligned} \tag{6.22}$$

The equation (6.22) can be further reparameterized in terms of Δ and a nuisance parameter. Next, similar to section 6.4.5, following Venzon and Moolgavkar [94], one can compute the profile likelihood-based confidence interval for different choices of the a. Pradhan et al. [72] investigated the performance of this interval for various choices of a and recommend $a = 2$ for use in practice.

The implementation of the above method in R is given below.

```
#Coupula package
library(copula)
#Profile likelihood method based on Venzon and Moolgavkar (1998)
library(Bhat)


binllk <- function(theta) {
  mycop<-frankCopula(1)
  pi1 <- theta[1]
```

```
  ## theta[2] = pi2 - pi1
  pi2 <- theta[1]-theta[2]
  mycop@parameters <- theta[3]

  if (pi1 >= 1 || pi1 <= 0) return (NA)
  if (pi2 >= 1 || pi2 <= 0) return (NA)
  #if (!copula:::chkParamBounds(mycop)) return (NA)
  p4 <- pCopula(c(1-pi1, 1-pi2), mycop)
  p2 <- pCopula(c(1, 1-pi2), mycop) - pCopula(c(1-pi1, 1-pi2), mycop)
  p3 <- pCopula(c(1-pi1, 1), mycop)-pCopula(c(1-pi1, 1-pi2), mycop)
  p1 <- pCopula(c(1, 1), mycop) - pCopula(c(1-pi1, 1), mycop) - pCopula(c(1, 1-pi2), mycop)
  + pCopula(c(1-pi1, 1-pi2), mycop)

  llk <- sum(dat * log(c(p1, p2, p3, p4)))
  return(-llk)
}
#adjustment, adjusting by adding 0.5 to each cell
a<-0.5

#enter data C(a,b,c,d), a=n11 b=n10 c=n01 d=n00
dat <- c(1+a,14+a,6+a,4+a)

#initialization of the parameters
p1.0 <-(dat[1]+dat[2]) / sum(dat)
p2.0 <-(dat[1]+dat[3]) / sum(dat)
del.0 <- p1.0 - p2.0
ifelse((del.0>0 || del.0<=1e-4),1e-4,del.0)
theta.0 <- c(p1.0, del.0, 2)

#Getting MLE of the parameters and the hessian matrix
result <- optim(theta.0, binllk, hessian=TRUE)
vc<-solve(result$hessian)

#Standard error calculation
se<-sqrt(diag(vc))
#-------Wald CI based on copula---------------------------------
c(result$par[2] -qnorm(0.9)*se[2], result$par[2] +qnorm(0.9)*se[2])

#Profile likelihood based estimates
mle<-result$par
x <- list(label=c("p1","dlt","alph"),est=mle,low=c(0.001,-0.99,-99),upp=c(0.99,0.99,99))
plkhci(x,binllk,"dlt",prob=0.9, nmax=10)
```

Using the same set-up as considered for drawing Figure 6.1, we plot the attained coverage probabilities of the copula-based CI using the profile likelihood method in Figure 6.6. Note that, in terms of attaining the 95% nominal level, the copula-based interval is a much-improved version of the Wald interval. There are few p_1's near the 0 and 1 for which the method shows slight undercoverages; otherwise, for all other points in the p_1-space, the method attains the nominal 95% level.

Illustration: Copula-based confidence interval

Consider the SpyGlass study example (Table 6.2) stated in section 6.4.1. The null hypothesis of interest is whether the probability of detecting malignancy using SpyGlass probe is less than or equal to that of the conventional biopsy. To test the hypothesis, we run the following R code (part of it is already included before but repeated here for the sake of completeness) for 90% CI.

```
#adjustment, adjusting by adding 0.5 to each cell
```

FIGURE 6.6
Coverage probabilities of copula-based confidence intervals for $n = 10$ with $p_2 = 0.5$ and $\phi = 1$.

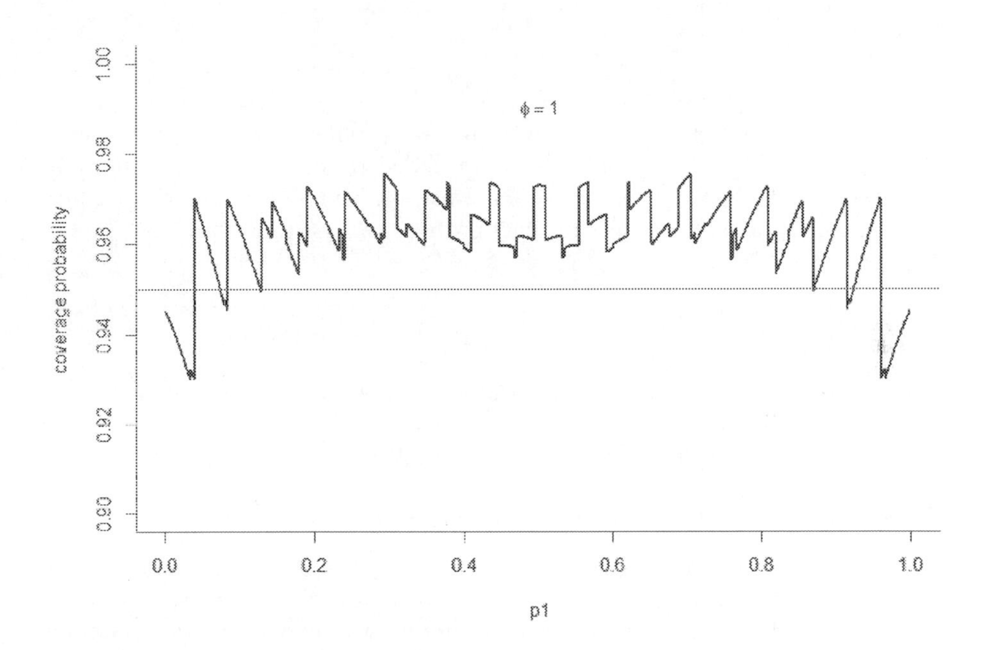


```
a<-0.5

#enter data C(a,b,c,d), a=n11 b=n10 c=n01 d=n00
dat <- c(1+a,14+a,6+a,4+a)

#initialization of the parameters
p1.0 <-(dat[1]+dat[2]) / sum(dat)
p2.0 <-(dat[1]+dat[3]) / sum(dat)
del.0 <- p1.0 - p2.0
ifelse((del.0>0 || del.0<=1e-4),1e-4,del.0)
theta.0 <- c(p1.0, del.0, 2)

#Getting MLE of the parameters and the hessian matrix
result <- optim(theta.0, binllk, hessian=TRUE)
vc<-solve(result$hessian)

#Standard error calculation
se<-sqrt(diag(vc))
#-------Wald CI based on copula---------------------------------
c(result$par[2] -qnorm(0.9)*se[2], result$par[2] +qnorm(0.9)*se[2])

#Profile likelihood based estimates
mle<-result$par
x <- list(label=c("p1","dlt","alph"),est=mle,low=c(0.001,-0.99,-99),upp=c(0.99,0.99,99))
plkhci(x,binllk,"dlt",prob=0.9, nmax=10)
```

The R output:

```
#-------Wald CI based on copula---------------------------------
> c(result$par[2] -qnorm(0.9)*se[2], result$par[2] +qnorm(0.9)*se[2])
[1] 0.09151479 0.50123147
...
#Profile likelihood based estimates
[1] 0.02329832 0.53803076
Warning messages:......
```

Note that the 90% CI due to Wald is $(0.09151479, 0.50123147)$ and the same using copula-based weighted profile likelihood method is $(0.02329832, 0.53803076)$. Since zero is excluded from the intervals, the null hypothesis is rejected by both these intervals, confirming that the probability of detecting malignancy using SpyGlass probe is not less than the conventional biopsy method.

6.4.7 Bayesian credible intervals

This section briefly discusses the Bayesian method for the difference of matched binomial proportions. As stated before, for $\boldsymbol{\theta} \equiv (p_{1+}, p_{+1}) \equiv (p_1, p_2)$, the Bayesian approach to the inference of $\boldsymbol{\theta}$ follows:

$$P(\boldsymbol{\theta}|\boldsymbol{x}) \propto P(\boldsymbol{x}|\boldsymbol{\theta})P(\boldsymbol{\theta}) \tag{6.23}$$

where $P(\boldsymbol{\theta}|\boldsymbol{x})$ is called posterior distribution, $P(\boldsymbol{x}|\boldsymbol{\theta})$ is the likelihood of the data, and $P(\boldsymbol{\theta})$ is a prior distribution of $\boldsymbol{\theta}$. In the Bayesian framework, posterior distribution gives the formal way of updating the prior knowledge about the possible values of the parameter $\boldsymbol{\theta}$ in light of the evidence provided by the observed data. The computation of the Bayesian interval (often referred to as the credible interval) is based on the posterior distribution. In chapter 1, we have included a brief review of Bayesian credible interval. We also recommend section 9.2.4 in Casella and Berger (2002) for further discussion on this topic.

A typical paired binomial table can be represented as table 6.1. The likelihood $L(\boldsymbol{p}|\boldsymbol{n}) \equiv P(\boldsymbol{x}|\boldsymbol{\theta})$ can be written as a multinomial distribution

$$L(\boldsymbol{p}|\boldsymbol{n}) = \frac{n!}{n_{11}!n_{10}!n_{01}!n_{00}!} p_{11}^{n_{11}} p_{10}^{n_{10}} p_{01}^{n_{01}} p_{00}^{n_{00}} \tag{6.24}$$

Since the maximum likelihood estimate of Δ, $\widehat{\Delta} = (n_{10} - n_{01})/n$, depends on the off-diagonal cell counts of 6.1 and the total n, one can reformulate the above multinomial as trinomial distribution as the following.

$$L(n_{10}, n_{01}|\Delta, p_{01}) = \frac{n!}{n_{10}!n_{01}!(n - n_{10} - n_{01})!}(p_{10})^{n_{10}} p_{01}^{n_{01}} (1 - p_{01} - p_{10})^{n - n_{10} - n_{01}}$$

These two formulations of the multinomial likelihood are equivalent. Therefore, WLOG we consider equation 6.24for our posterior computation. As discussed earlier, one of the most important aspects of Bayesian analysis is the choice of a prior distribution $P(\boldsymbol{\theta})$. For multinomial likelihood, the *Diritchlet*$(\boldsymbol{\alpha})$ with $\boldsymbol{\alpha} \equiv (\alpha_1, \alpha_2, \alpha_3, \alpha_4)$ is the conjugate prior for $\boldsymbol{\theta}$. The *pdf* of the *Diritchlet*$(\boldsymbol{\alpha})$ prior given by

$$Dirichlet(\boldsymbol{\theta}|\boldsymbol{\alpha}) = \frac{1}{C(\boldsymbol{\alpha})} p_{11}^{(\alpha_1-1)} p_{10}^{(\alpha_2-1)} p_{01}^{(\alpha_3-1)} p_{00}^{(\alpha_4-1)}$$

where $C(\boldsymbol{\alpha})$ is the normalizing constant. A conjugate prior is a prior distribution where prior and the posterior belong to the same family of distributions. For paired Binomials, using *Dirichlet*$(\boldsymbol{\alpha})$ as the conjugate prior for the likelihood $L(\boldsymbol{p}|\boldsymbol{n})$, the posterior (equation 6.23) becomes

$$\begin{aligned} P(\boldsymbol{\theta}|\boldsymbol{x}) &\propto P(\boldsymbol{x}|\boldsymbol{\theta})P(\boldsymbol{\theta}) \\ &\propto L(\boldsymbol{p}|\boldsymbol{n}) \times Dirichlet(\boldsymbol{\alpha}) \\ &\propto p_{11}^{n_{11}} p_{10}^{n_{10}} p_{01}^{n_{01}} p_{00}^{n_{00}} \times p_{11}^{(\alpha_1-1)} p_{10}^{(\alpha_2-1)} p_{01}^{(\alpha_3-1)} p_{00}^{(\alpha_4-1)} \\ &\propto p_{11}^{(n_{11}+\alpha_1-1)} p_{10}^{(n_{10}+\alpha_2-1)} p_{01}^{(n_{01}+\alpha_3-1)} p_{00}^{(n_{00}+\alpha_4-1)} \\ &\propto Dirichlet(\boldsymbol{n}+\boldsymbol{\alpha}) \end{aligned} \tag{6.25}$$

Note that the Dirichlet parameters $\boldsymbol{\alpha}$ can be regarded as the "pseudo-counts" from "pseudo-data".
In Bayesian inference, often Markov Chain Monte Carlo (MCMC) methods are used to generate samples from a posterior distribution. In this set up, using Gibbs sampling approach, which is a form of MCMC, one can easily generate the marginal distribution samples $p_{ij}^1, p_{ij}^2, .., p_{ij}^N$ for the parameter p_{ij}, $i, j = 1, 0$. In turn, one can generate samples $p_{10}^1 - p_{01}^1, p_{10}^2 - p_{01}^2, .., p_{10}^N - p_{01}^N$ for the marginal distributions of $p_{10} - p_{01}$. The $100 \times (1-\alpha)\%$ credibility interval of $p_{10} - p_{01}$ is computed using the sample percentiles of the generated values that induce the shortest credible intervals. Note that for $\alpha_1 = \alpha_2 = \alpha_3 = \alpha_4 = 1$, the Dirichlet distribution is equivalent to a uniform distribution over all points in its support, and $\alpha_1 = \alpha_2 = \alpha_3 = \alpha_4 = 0.5$ is the Jeffreys non-informative prior.

Since the posterior distribution is the joint distribution of four parameters, namely, p_{11}, p_{10}, p_{01}, and p_{00}, the main point of interest here is to generate samples from the four marginal distributions of these parameters. Once the samples from the marginals for these parameters are obtained, the computation of the credible interval of any function of parameters is straightforward. For example, if the primary interest is to compute the interval for the difference of proportions $\Delta = p_{10} - p_{01}$, then the following PROC MCMC can be run to accomplish the job. Consider the registry study data using SpyGlass Direct Visualization system data example given in Pradhan et al. [72]. In this study, the diagnosis of malignancy was carried out on 25 patients using SpyGlass and conventional biopsy. The null hypothesis of interest was whether the probability of detection of malignancy using SpyGlass probe is less than that of the conventional biopsy. The data are given in the following in Table 6.4.

TABLE 6.4
SpyGlass ERCP versus conventional biopsy results

		Biopsy results		
		Benign	**Malignant**	
SpyGlass results	**Benign**	1	14	15
	Malignant	6	4	10
		7	18	25

We are interested in computing the 90% credible intervals using Bayesian framework. First, we enter data $\boldsymbol{n} = (n_{11}, n_{10}, n_{01}, n_{00}) = (1, 14, 6, 4)$ using the datastep given below and then run PROC MCMC using the following syntax:

```
data SpyGlass;
   input n11 n10 n01 n00;
datalines;
1 14 6 4
;

ods graphics on;
proc mcmc data=SpyGlass seed=12345 outpost=o1 nmc=100000 nthin=10 monitor=(delta)
stats(alpha=(0.1))=(int summary) ;
   array n[4] n11 n10 n01 n00;
   array p[4];
   array a[4] (0.5 0.5 0.5 0.5) ;

   parm p;
   prior p ~ dirichlet(a);
   model n ~ multinomial(p);
   beginnodata;
delta=p2-p3;
   endnodata;
   run;
ods graphics off;
```

In the above PROC MCMC code, *n*mc=100000 invokes the number of simulations to be generated, and *n*thin=10 option invokes the thinning to be applied on the samplings. In the PROC MCMC code, the *array p[4] p1 p2 p3 p4* defines the four parameters p_1 and p_2, p_3, p_4. The option *p*arms specify the parameters for which marginals to be computed. The option *p dirichlet(a)* specifies the priors $Dirichlet(0.5, 0.5, 0.5, 0.5)$ corresponding to the 4 p_1, p_2, p_3, p_4 parameters. The option *model n multinomial(p)* invokes likelihood as multinomial with parameter $\boldsymbol{p}$. Finally, the command *delta=p1-p2* within *beginnodata* and *endnodata* computes the difference *delta* of the two marginals, and the option *monitor=(delta)* displays the output related to *delta*.

Figure 6.7 shows the diagnostics for "delta", which is derived from the marginals of the parameters p_{10} and p_{01}. Since the trace plot of "delta" shows good mixing and other diagnostic plots show no apparent issues, one can assume that all chains have converged. The plots corresponding to the Autocorrelation panel show a small degree of autocorrelation present at the beginning of the sampling, and then it quickly diminishes.

SAS's PROC MCMC shows the following as the posterior summary.

Posterior Summaries

Parameter	N	Mean	Standard Deviation	Percentiles 25	50	75
delta	10000	0.2977	0.1570	0.1964	0.3028	0.4098

Posterior Intervals

Parameter	Alpha	Equal-Tail Interval		HPD Interval	
delta	0.100	0.0263	0.5435	0.0384	0.5527

In the above, SAS output includes the *Posterior Summaries* and *Posterior Intervals*. The *Posterior Summaries* include the posterior summary of the parameter "delta". The mean, standard deviations, and 25th, 50th, and 75th percentiles are shown as 0.2977, 0.157, 0.1964, 0.3028, and 0.4008, respectively. Just below the *Posterior Summaries* part, the outputs corresponding to the posterior intervals are shown in *Posterior Intervals*. The 90% HPD interval (highest posterior density interval) for delta is shown as $(0.0384, 0.5527)$, and the 90% equal tail interval for delta is shown as $(0.0263, 0.5435)$. Note that both the HPD interval and the equal tail interval exclude 0, which indicates the detection rates of using these two different systems are different.

Once the samples from the marginal distributions for the parameters $p_1 == p_{1+} = p_{11} + p_{10}$ and $p_2 \equiv p_{+1} = p_{11} + p_{01}$ are generated, one can easily compute a credible interval for any statistic. For example, if one is interested in generating the interval for the ratio of proportions p_1 and p_2 (i.e., $\frac{p_1}{p_2}$ with priors $Dirichlet(0.5, 0.5, 0.5, 0.5)$, then the following code can be run.

```
data SpyGlass;
   input n11 n10 n01 n00;
datalines;
1 14 6 4
;

ods graphics on;
proc mcmc data=SpyGlass seed=12345 outpost=o1 nmc=100000 nthin=10 monitor=(Rho)
stats(alpha=(0.1))=(int summary) ;
   array n[4] n11 n10 n01 n00;
   array p[4];
   array a[4] (0.5 0.5 0.5 0.5) ;
```

```
   parm p;
   prior p ~ dirichlet(a);
   model n ~ multinomial(p);
   beginnodata;
Rho=(p1+p2)/(p1+p3);
   endnodata;
   run;
ods graphics off;
```

The following diagnostic plots shown in Figure 6.7 from PROC MCMC indicate no issue with the Markov chains, and the posterior summary for the ratio of proportions ϕ shows an HPD interval as $(0.6011, 1.1774)$. For details about the diagnostic plots we refer to the discussion in page 97

FIGURE 6.7
Bayesian diagnostics and posterior summaries using PROC MCMC for p_1/p_2.

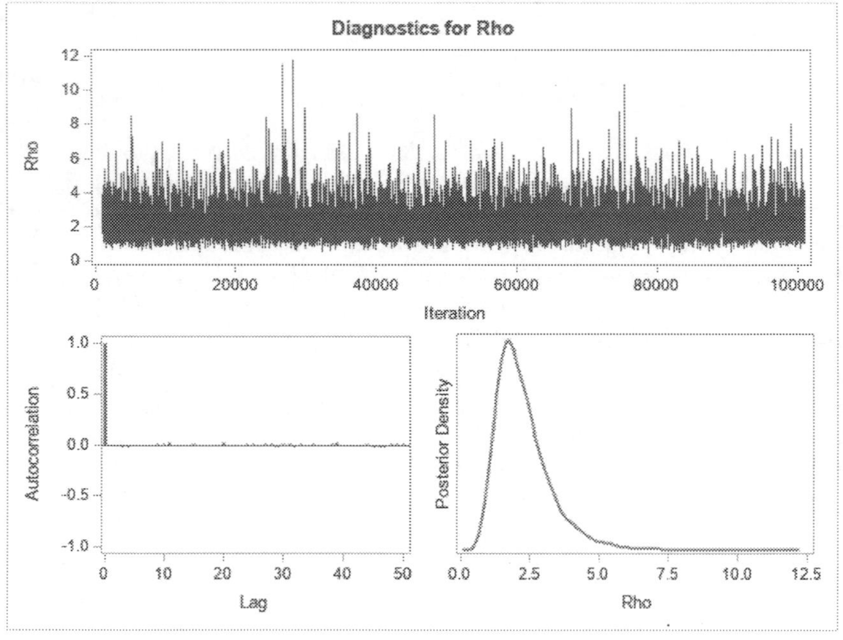


```
                    Posterior Summaries and Intervals

                                        Standard
Parameter          N        Mean       Deviation       95% HPD Interval

phi            10000      0.8919          0.1493       0.6011        1.1774
```

6.5 Exact confidence intervals

In a small sample setup, exact confidence intervals are recommended. The exact methods are designed to ensure the attainment of nominal coverage probability. However, this advantage of the exact confidence intervals comes at a cost. The confidence intervals are usually conservative (wider) and computationally challenging even with moderate sample sizes. The following section presents an unconditional exact method for paired binomial data due to Sidik [84].

6.5.1 Exact interval by Sidik

The unconditional exact method proposed by Sidik [84] utilizes the score statistic. As stated before, under null hypothesis let $\Delta = p_{1+} - p_{+1} = \Delta_0$, then for $\Delta_0 \in [-1, 1]$ the score statistic is defined as the following.

$$T(\boldsymbol{n}|\Delta_0) = \frac{n_{10} - n_{01} - n\Delta_0}{\sqrt{n[2\tilde{p}_{01} + \Delta_0(1 - \Delta_0)]}}$$

where $\boldsymbol{n} = (n_{11}, n_{10}, n_{01}, n_{00})$ is the observed table of type Table 6.1 with $n = n_{11} + n_{10} + n_{01} + n_{00}$, $\tilde{p}_{01}$ is the maximum likelihood estimate of p_{01} under the constraint $p_{1+} - p_{+1} = \Delta_0$. The estimate $\tilde{p}_{01}$ can be derived as

$$\tilde{p}_{01} = \frac{-b + \sqrt{b^2 - 4ac}}{2a}$$

with $a = 2n$, $b = (2n + n_{10} - n_{01})_0 - n_{10} - n_{01}$ and $c = -n_{01}\Delta_0(1 - \Delta_0)$. Let $\boldsymbol{x} = (x_{11}, x_{10}, x_{01}, x_{00})$ be a generic table. Since the estimate of $\Delta = p_{1+} - p_{+1}$ is $(x_{10} - x_{01})/n$ and it depends only on the off-diagonal elements of the observed frequency, the probability mass function of a generic table $\boldsymbol{x}$ can be written as

$$\begin{aligned} f(x_{10}, x_{01}|p_{10}, \Delta_0, n) &= \frac{n!}{x_{10}!x_{01}!(n - x_{10} - x_{01})!} \\ &\quad p_{10}^{x_{10}}(p_{10} - \Delta_0)^{x_{01}}(1 - 2p_{10} + \Delta_0)^{n - x_{10} - x_{01}} \end{aligned} \tag{6.26}$$

where $\Delta_0 = p_{10} - p_{01}$. Clearly, in equation 6.26, the parameter Δ_0 is the parameter of interest, and the parameter p_{10} is the nuisance parameter with the domain $0 \leq p_{10} \leq (1 - \Delta_0)/2$. The 100(1-$\alpha$)% confidence interval $(\underline{\Delta}, \overline{\Delta})$ is obtained iteratively using the following.

$$\text{Sup}_{p_{10}} \left\{ \sum_{\omega(x|\Delta_0,n)} I[T(\boldsymbol{x}|\underline{\Delta}) \geq T(\boldsymbol{n}|\underline{\Delta}) f(x_{10}, x_{01}|p_{10}, \underline{\Delta}, n) \right\} = \alpha/2$$

$$\text{Sup}_{p_{10}} \left\{ \sum_{\omega(x|\Delta_0,n)} I[T(\boldsymbol{x}|\overline{\Delta}) \leq T(\boldsymbol{n}|\overline{\Delta}) f(x_{10}, x_{01}|p_{10}, \overline{\Delta}, n) \right\} = \alpha/2$$

Using the same set-up as considered for Figure 6.1, we plot the attained coverage probabilities of the exact interval due to Sidik in Figure 6.8. Notice that, the exact interval always attained the 95% nominal level.

FIGURE 6.8
Coverage probabilities of unconditional exact confidence intervals for $n = 10$ with $p_2 = 0.5$ and $\phi = 1$.

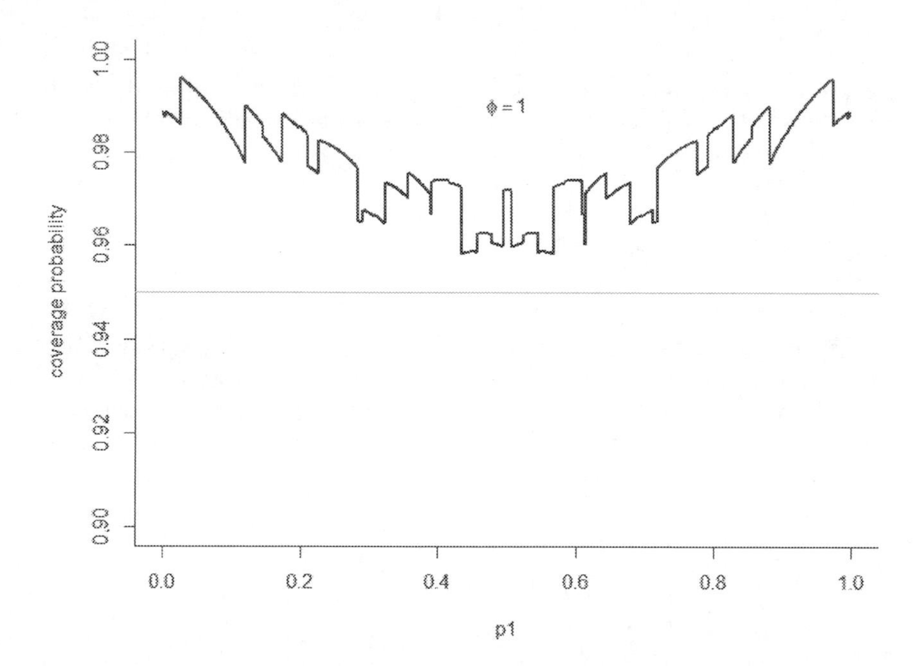

Illustration: Tango interval and Sidik interval

Consider the SpyGlass study example (Table 6.2) stated in section 6.4.1. The null hypothesis of interest is whether the probability of detecting malignancy using SpyGlass probe is less than equal to that of the conventional biopsy. In order to test the hypothesis, we run the following SAS code for 90% CI using the SAS version of StatXact, StatXact PROCs.

```
/*PROC BINOMIAL computes CI for the p2-p1, hence the input off-diagonal counts adjusted*/
Data Spyglass;
input ro col wgt;
cards;
1 1 1
1 2 6
2 1 14
2 2 4
;
/*poptype="rel" is the syntax for paired binomials--------------*/
/*alpha=0.9 request for 90% CI, default is 95% CI----------------*/
proc binomial data=Spyglass poptype="rel" out=out1 alpha=0.9;
pd/ex; /*syntax for diff of proportions */
pops ro -col;
weight wgt;
run;
```

The SAS output:

```
Output from StatXact (r) (v11.1) PROCs SAS9.4 (64-bit) TS1M5
Copyright (c) 1997-2018 Cytel Inc., Cambridge, MA, USA.
CONFIDENCE INTERVAL ON DIFFERENCE OF TWO RELATED BINOMIAL PROPORTIONS
BASED ON STANDARDIZED STATISTIC & INVERTING TWO 1-SIDED TESTS

Data file name : < SPYGLASS >
Population Variable 1 : ro
Population Variable 2 : col
Weight Variable Name : wgt


Statistics based on the observed  2 by  2 table :

    Observed proportion for population <        1> : piHat_1          =    0.2800
    Observed proportion for population <        2> : piHat_2          =    0.6000
    Observed difference of proportions : piHat_2-piHat_1              =    0.3200
    Stderr (pooled estimate of stdev of piHat_2-piHat_1)              =    0.1789
    Standardized difference (t): (piHat_2-piHat_1)/Stderr             =    1.7889

Results:
---------------------------------------------------------------------------------
                                P-value                      90.00% Conf.
Interval
    Method   1-sided(Pr{T .GE. t})     2*1-sided            for pi_2-pi_1
---------------------------------------------------------------------------------
      Asymp             0.0368               0.0736       (      0.0262,  0.5590)
      Exact             0.0518               0.1037       (     -0.0057,  0.5937)
```

In the above SAS output from PROC BINOMIAL shows the exact CI as $(-0.0057, 0.5937)$. Since the exact CI excludes zero, the null hypothesis is failed to be rejected. Note that the CI corresponding to the method **Asymp** shows an asymptotic confidence interval as $(0.0262, 0.5590)$. In StatXactPROCs's PROC BINOMIAL, asymptotic CI is computed using the score interval due to Tango. Interestingly, for this example, asymptotic CI excludes zero, but for the same using exact CI includes zero. This creates a conflicting statistical inference. Later in the discussion section, we have included a comparison for different methods using criterion expected coverage and expected lengths.

In computing the exact CI, we have used PROC BINOMIAL of StatXact PROCs. To our knowledge, this is the only software package where this method has been implemented. Note that in general, StatXact or StatXact PROCs is able to compute a CI with any combinations of the cell counts of a 2×2 table. The software, however, shows an incorrect error message when both of the off-diagonal cell counts (n_{10} and n_{01}) are zero. In those situations, we recommend computing a CI using Tango as stated in section 6.4.4.

6.6 Confidence intervals with missing data

Let Y_1 and Y_2 be two binary random variables. In a clinical trial setup, one might interpret Y_1 as the condition of the patient before the treatment and Y_2 after the treatment. Let us consider a data set where some of the subjects are fully observed, and hence the matched pairs can be created, whereas others are partially observed. The data can be presented as shown in Table 6.5. where

TABLE 6.5
Paired binomial table with missing pairs

$Y_1 \rightarrow$	y_{11}	y_{12}	$\cdot\cdot$	y_{1n}	$y_{1(n+1)}$	$\cdot\cdot$	$y_{1(n+l_1)}$			
$Y_2 \rightarrow$	y_{21}	y_{22}	$\cdot\cdot$	y_{2n}	$\cdots$		$\cdots$	$y_{2(n+1)}$	$\cdot\cdot$	$y_{2(n+l_2)}$

y_{ij} is either 1 or 0. The data can then be summarized as shown in Table 6.6.

TABLE 6.6
Paired binomial table with incomplete data

	$\boldsymbol{Y}_{2=1}$	$\boldsymbol{Y}_{2=0}$	**Missing on $\boldsymbol{Y}_1$**	**Total**
$\boldsymbol{Y}_{1=1}$	n_{11}	n_{10}	w_1	$n_{1+} + w_1$
$\boldsymbol{Y}_{1=0}$	n_{01}	n_{00}	$l_1 - w_1$	$n_{0+}l_1 - w_1$
Missing on $\boldsymbol{Y}_2$	w_2	$l_2 - w_2$		l_2
Total	$n_{+1} + w_2$	$n_{+0} + l_2 - w_2$	l_1	$N = n + l_1 + l_2$

Note the following:

- As in Table 6.1, we define $n_{ij}, n_{0+}, n_{1+}, n_{+0}$, and n_{+1} corresponding to the observations $(Y_1 = i, Y_2 = j), i, j = 0, 1$, and $n = n_{11}+n_{10}+n_{01}+n_{00}$. Thus, $\boldsymbol{n} \equiv (n_{11}, n_{10}, n_{01}, n_{00})'$ represents a random sample from a multinomial distribution with $\boldsymbol{p} \equiv (p_{11}, p_{10}, p_{01}, p_{00})$.

- Let $l_1(l_2)$ be the number observations where $Y_1(Y_2)$ is observed but not $Y_2(Y_1)$, and $w_1(w_2)$ is the number of Y_1 equal to 1 (Y_2 equal to 1) among the $l_1(l_2)$ $Y_1(Y_2)$ observations.

- Let $\delta = p_{+1} - p_{1+} = p_{01} - p_{10}$.

6.6.1 Confidence interval due to Chang (2011)

First, we discuss the construction of a confidence interval of δ with missing data due to Chang [23]. From Table 6.6, it is evident that w_1 and w_2 have $Bin(l_1, p_{1+})$ and $Bin(l_2, p_{+1})$ distributions, respectively, where $p_{1+} = p_{11}+p_{10}$ and $p_{+1} = p_{11} + p_{01}$. Thus, $\left(\frac{w_2}{l_2} - \frac{w_1}{l_1}\right)$, say, $\hat{\delta}_u$ is an unbiased estimator of δ provided l_1 and l_2 are both greater than 0. Also, $\frac{n_{01}-n_{10}}{n}$, say, $\hat{\delta}_p$ is another unbiased estimator of δ based on fully observed part of the data. Chang (2011) considers an estimator of the form $\delta_\alpha = \alpha\hat{\delta}_p + (1-\alpha)\hat{\delta}_u$, and chooses α such that it minimizes the variance of δ_α. Let $\hat{\alpha}$ be that value of α. With a little bit of algebra, one can find $\hat{\alpha} = \frac{\hat{B}}{\hat{A}+\hat{B}}$ where,

$$\begin{aligned}
\hat{A} = \quad & \mathrm{Var}(\hat{\delta}_p) = \left\{\frac{n_{01}}{n}\left(1 - \frac{n_{01}}{n}\right) + \frac{n_{10}}{n}\left(1 - \frac{n_{10}}{n}\right) + 2\frac{n_{10}n_{01}}{n^2}\right\}\frac{1}{n} \\
\hat{B} = \quad & \mathrm{Var}(\hat{\delta}_u) = \frac{1}{l_2}\frac{w_2}{l_2}\left(1 - \frac{w_2}{l_2}\right) + \frac{1}{l_1}\frac{w_1}{l_1}\left(1 - \frac{w_1}{l_1}\right)
\end{aligned}$$

Let us write $\hat{\delta}_\alpha = \hat{\alpha}\hat{\delta}_p + (1-\hat{\alpha})\hat{\delta}_u$. Finally, the $100(1-\alpha)\%$ Wald type confidence interval of δ based on $\hat{\delta}_\alpha$ is given by

$$\frac{\hat{B}}{\hat{A}+\hat{B}}\hat{\delta}_p + \frac{\hat{A}}{\hat{A}+\hat{B}}\hat{\delta}_u \pm z_{\alpha/2}\sqrt{(\hat{A}\hat{B})/(\hat{A}+\hat{B})}, \tag{6.27}$$

where $z_{\alpha/2}$ is the upper 100α percentile point of the standard normal distribution.

The SAS implementation of the method proposed by Chang is given in the following:

```
%macro chang(n11= , n10= ,n01= ,n00= , l1= ,w1= ,l2= ,w2= ,prob=0.05);
data _xx;
n11=&n11;
n12=&n10;
n21=&n01;
n22=&n00;
prob=&prob;
w1=&w1; w2=&w2;l1=&l1; l2=&l2;
n=n11+n12+n21+n22;
delta_phat=(n12-n21)/n;
_A=n21*(n-n21)/n**3 +n12*(n-n12)/n**3 +(2*n21*n12)/n**3;
delta_uhat=w2/l2-w1/l1;
_B=w2*(l2-w2)/l2**3 + w1*(l1-w1)/l1**3 ;
alpha=_B/(_A+_B);
delta_hat=alpha*delta_phat+(1-alpha)*delta_uhat;
var_deltahat=(_A*_B)/(_A+_B);
```

```
Chang_lo=delta_hat-probit(1. - prob/2)*sqrt(var_deltahat);
Chang_up=delta_hat+probit(1. - prob/2)*sqrt(var_deltahat);
keep Chang_lo Chang_up;
run;
options nodate nonumber;
title " Confidence interval of paired binomial proportions with incomplete pairs";
proc print data=_xx noobs ;
run;

%mend;
```

Illustration: Chang (2011) confidence interval

Consider the neurological data from Choi and Stablein[24] on 33 young meningitis participants. According to the study, $n_{00} = 6$, $n_{01} = 3$, $n_{10} = 8$, $n_{11} = 8$, $n = 25$, $w_1 = 4$, $l_1 = 6$, $w_2 = 2$, and $l_2 = 2$. Some observations are assumed to be missing due to the administrative reasons, and hence may be assumed to be missing at random (MAR). The following SAS code can be run to compute the 95% CI of the difference of proportions.

```
/*Chang (2011), CI of the difference of proportions of the paired binomial with incomplete
 pairs*/
%chang(n11=8 , n10=8 ,n01=3 ,n00=6 , l1=6 ,w1=4 ,l2=2 ,w2=2 ,prob=0.05);
```

The SAS output:

```
Confidence interval of paired binomial proportions with incomplete pairs

                         Chang_lo     Chang_up

                         0.033049      0.44740
```

Clearly, the 95% CI is $(0.033049, 0.44740)$. Since the CI excludes zero, the null hypothesis is rejected at 5% level of significance, and confirms that the neurological complication rates before and after treatment are plausibly different.

6.6.2 Likelihood-based confidence intervals

Following Tang et al. [90], Pradhan et al. [71] propose four intervals for $\delta = p_{1+} - p_{+1} = p_{10} - p_{01}$, the difference between two success rates. The two of these intervals are likelihood-based Wald type, and the two are profile likelihood-based. In the following, we discuss these intervals.

6.6.2.1 Likelihood-based Wald-type intervals

Consider the data described in Table 6.6. Note that one can rewrite $p_{10} = p_{01} + \delta$, $p_{11} = 1 - (2p_{01} + p_{00} + \delta)$, $p_{1+} = 1 - p_{01} - p_{00}$ and $p_{+1} = 1 - p_{01} - p_{00} - \delta$. Thus, the joint probability distribution of $(n_{11}, n_{10}, n_{01}, n_{00})$ in terms of δ, p_{01}, p_{00} is given by,

$$P(\boldsymbol{n}|\delta, p_{01}, p_{00}) = \frac{n!}{n_{11}!n_{10}!n_{01}!n_{00}!}(p_{01}+\delta)^{n_{10}} \times {p_{01}}^{n_{01}}{p_{00}}^{n_{00}}(1-2p_{01}-p_{00}-\delta)^{n_{11}}, \tag{6.28}$$

and that of (w_1, w_2) is given by,

$$P(w_1, w_2|\delta, p_{01}, p_{00}) = \frac{l_1}{w_1!(l_1-w_1)!}(1-p_{01}-p_{00})^{w_1}(p_{01}+p_{00})^{(l_1-w_1)} \times \frac{l_2}{w_2!(l_2-w_2)!}(1-p_{01}-p_{00}-\delta)^{w_2}(p_{01}+p_{00}+\delta)^{(l_2-w_2)}. \tag{6.29}$$

Combining (6.28) and (6.29), we obtain the likelihood function of $\boldsymbol{\theta} = (\delta, p_{01}, p_{00})$ based on the data in Table 6.6 as,

$$\begin{aligned} P(\boldsymbol{\theta}|\boldsymbol{n}, w_1, w_2) = & \frac{n!}{n_{11}!n_{10}!n_{01}!n_{00}!}(p_{01}+\delta)^{n_{10}} \\ & \times {p_{01}}^{n_{01}}{p_{00}}^{n_{00}}(1-2p_{01}-p_{00}-\delta)^{n_{11}} \\ & \times \frac{l_1}{w_1!(l_1-w_1)!}(1-p_{01}-p_{00})^{w_1}(p_{01}+p_{00})^{(l_1-w_1)} \\ & \times \frac{l_2}{w_2!(l_2-w_2)!}(1-p_{01}-p_{00}-\delta)^{w_2}(p_{01}+p_{00}+\delta)^{(l_2-w_2)} \end{aligned} \tag{6.30}$$

Notice that, δ is our parameter of interest and rest are nuisance parameters.

Suppose $\widehat{\delta}$ is the maximum likelihood estimate of the parameter of interest δ and $\widehat{se}(\hat{\delta})$ is an estimate of its standard error, then the $100(1-\alpha)\%$ Wald confidence interval is given by $(\widehat{\delta} - z_{\alpha/2}\widehat{se}(\hat{\delta}), \widehat{\delta} + Z_{\alpha/2}\widehat{se}(\hat{\delta}))$ where $z_{\alpha/2}$ is the $100(1-\alpha/2)$ upper percentile of the standard normal distribution.

We call this interval Tang-Wald or **TW** interval. Next, applying Agresti and Min [5] correction to the multinomial part, and Agresti and Coull [3] adjustment to the binomial part of the likelihood the corrected likelihood becomes

$$\begin{aligned} P_C(\boldsymbol{\theta}|\boldsymbol{n}, w_1, w_2) \propto \; & (p_{01}+\delta)^{(n_{10}+0.5)}{p_{01}}^{(n_{01}+0.5)}{p_{00}}^{(n_{00}+0.5)} \\ & (1-2p_{01}-p_{00}-\delta)^{(n_{11}+0.5)} \\ & \times(1-p_{01}-p_{00})^{w_1+1}(p_{01}+p_{00})^{(l_1-w_1-1)} \\ & \times(1-p_{01}-p_{00}-\delta)^{w_2+1}(p_{01}+p_{00}+\delta)^{(l_2-w_2-1)}. \end{aligned} \tag{6.31}$$

Similar to the **TW** interval, the $100(1-\alpha)\%$ confidence interval of δ based on the maximum corrected likelihood estimate of δ can be obtained. Since the likelihood involves a correction, we name this method as **TWC**.

6.6.2.2 Profile likelihood-based confidence interval

Pradhan, et al. [71] propose two profile likelihood-based confidence intervals using the likelihood functions $P(\boldsymbol{\theta}|\boldsymbol{n}, w_1, w_2)$ (cf.(6.30)), and $P_C(\boldsymbol{\theta}|\boldsymbol{n}, w_1, w_2)$ (cf.(6.31)). Let $\boldsymbol{\theta} = (\delta, p_{01}, p_{00})$, and δ be the parameter of interest. Following the work of Venzon and Moolgavkar [94], we utilize the profile likelihood approach to construct a confidence interval for δ. Let $\Theta = \{\boldsymbol{\theta}| -1 \leq \delta \leq 1, 0 \leq p_{01} \leq 1, 0 \leq p_{00} \leq 1\}$ denote the parameter space. Assuming δ to be fixed, we maximize $\ell(\Theta) = \ell(\delta, p_{01}, p_{00})$ with respect to p_{01} and p_{00}. The value of p_{01} and p_{00} maximizing $\ell(\Theta)$ for given δ will naturally be in terms of δ, say, $p_{01}(\delta)$ and $p_{00}(\delta)$. Then the maximum profile likelihood function of δ, denoted by $\tilde{\ell}(\delta)$ is given,

$$\tilde{\ell}(\delta) = \ell(\delta, p_{01}(\delta), p_{00}(\delta)) \tag{6.32}$$

Then, the $100(1-\alpha)\%$ profile likelihood-based confidence interval for δ is given by

$$\left\{\delta : \ell(\widehat{\boldsymbol{\theta}}) - \tilde{\ell}(\delta) \leq \frac{1}{2}\chi_1^2(1-\alpha)\right\}, \tag{6.33}$$

where $\widehat{\boldsymbol{\theta}}$ is the maximum likelihood estimate of $\boldsymbol{\theta}$ as defined above, and $\chi^2_{1(1-\alpha)}$ is the $100(1-\alpha)$ percentile of χ_1^2 distribution.

Following Venzon and Moolgavkar [94], the confidence interval for δ, say, (δ_L, δ_U), is the admissible solution to the following system of non-linear equations:

$$\begin{bmatrix} l(\hat{\boldsymbol{\theta}}) - l(\boldsymbol{\theta}) - \frac{1}{2}\chi_1^2(\alpha) \\ \frac{\partial l(\delta, p_{01}, p_{00})}{\partial p_{01}} \\ \frac{\partial l(\delta, p_{01}, p_{00})}{\partial p_{00}} \end{bmatrix} = \mathbf{0}.$$

Since the interval uses the profile likelihood method with the likelihood proposed by Tang et al. [89], we name this method as **TPF**. Finally, similar to the **TPF** method, by incorporating equation (6.31), instead of equation (6.30), one can compute the corrected profile likelihood confidence interval using the corrections suggested by Agresti and Min [5] and Agresti and Coull [3], we name this as **TPFC**.

For the paired binomial set-up, Agresti and Min [5] pointed out that there is no definite justification for the choice of cell weight because of its ad hoc nature. In the weighted profile likelihood method, following Agresti and Coull [3], Pradhan et al. [71] recommended adding 0.5 to each cell of the 2×2 table and adding 1 to each cell of the binomial part of likelihood.

The following is the SAS/IML implementation of the above method.

```
%macro multinomPF (n00= , n01= ,n10= ,n11= , u= ,m1= ,v= ,m2=, adj=,adju=, adjv= ,
%alpha=);
proc iml;
Start likelihood(x)global(n00,n01,n10,n11,u,v,m1,m2);
L=n10*log(x[2]+x[3])+n01*log(x[2])+n11*log(x[1])+n00*log(1-x[1]-2*x[2]-x[3])
+u*log(x[1]+x[2]+x[3])+(m1-u)*log(1-x[1]-x[2]-x[3])
+v*log(x[1]+x[2])+(m2-v)*log(1-x[1]-x[2]);
/* print L;*/
return(L);
finish likelihood;

start Gradient(x)global(n00,n01,n10,n11,u,v,m1,m2);
g=j(1,3,1e-16);
g[1]=n11/x[1] - n00/(1-x[1]-2*x[2]-x[3])+u/(x[1]+x[2]+x[3])-(m1-u)/(1-x[1]-x[2]-x[3])
      +v/(x[1]+x[2])-(m2-v)/(1-x[1]-x[2]);
g[2]=n10/(x[2]+x[3])+n01/x[2] -(2*n00)/(1-x[1]-2*x[2]-x[3])+u/(x[1]+x[2]+x[3])
      -(m1-u)/(1-x[1]-x[2]-x[3])+v/(x[1]+x[2])-(m2-v)/(1-x[1]-x[2]);
g[3]=n10/(x[2]+x[3]) -n00/(1-x[1]-2*x[2]-x[3]) +u/(x[1]+x[2]+x[3]) -(m1-u)/(1-x[1]-x[2]-
x[3]); return (g);
finish Gradient;

 x0 ={0.05 0.05 1e-4};
 n00=%sysevalf(&n00+&adj);n01=%sysevalf(&n01+&adj);n10=%sysevalf(&n10+&adj);n11=
 %sysevalf(&n11+&adj);
u=%sysevalf(&u+&adju);v=%sysevalf(&v+&adjv);m1=%sysevalf(&m1+2*&adju);
 m2=%sysevalf(&m2+2*&adjv);

 optn = {1 0};
 con = { 1.e-4 1.e-4 -0.999 . .,
  0.999 0.999 0.999 . . ,
. 1 1 1 1e-4,
1 2 1 -1 0.99,
1 1 1 1 1e-4,
1 1 1 -1 0.99,
1 1 . -1 0.99};
 call nlptr(rc,xres,"likelihood",x0,optn,con,,,,"Gradient");
 xopt = xres`; fopt = likelihood(xopt);

  call nlpfdd(f,g,hes2,"likelihood",xopt,,"Gradient");
/* computing wald CI*/
hin2 = inv(hes2);
   /* quantile of normal distribution */
   prob = .05;
   noqua = probit(1. - prob/2);
   stderr = sqrt(abs(vecdiag(hin2)));
   xlbw = xopt - noqua * stderr;
   xubw = xopt + noqua * stderr;
ci=xlbw||xubw;
create WldCI from ci[colname={'Lower','Upper'} ] ;
append from ci;
close WldCI;


/* print hes2;*/
 start plgrad(x) global(like,ipar,lstar);
 like = likelihood(x);
 grad = Gradient(x);
 grad[ipar]=like-lstar;
/* print grad;*/
return(grad`);
 finish plgrad;
```

```
prob=&alpha. ;
  xlb={1e-4, 1e-4, -0.999};
xub=j(3,1,1);
/* quantile of chi**2 distribution */
chqua = cinv(1-prob,1);like=fopt; lstar = fopt - .5 * chqua;

 optn = {3 0};
 do ipar = 3 to 3;
/*  Implementation of Venzon & Moolgavkar (1988)*/

delt = - inv(hes2[1:2,1:2]) * hes2[1:2,3];

alfa = - (hes2[3,3] - delt` * hes2[1:2,3]);
if alfa > 0 then alfa = .5 * sqrt(chqua / alfa);
else do;
print "Bad alpha";
alfa = .1 * xopt[ipar];
end;
if ipar=1 then delt = {1}// delt;
else delt = delt //{1};
/* print delt;*/
x0 = xopt + (alfa * delt);
con2 = con; con2[1,ipar] = xopt[ipar];
tc={2000,5000};
call nlplm(rc,xres,"plgrad",x0,optn,con2,tc );

f = plgrad(xres); s = ssq(f);
if (s <1.e-6) then xub[ipar] = xres[ipar];
else xub[ipar] =1.0;
x0 = xopt - (alfa * delt);
con2[1,ipar] = con[1,ipar]; con2[2,ipar] = xopt[ipar];
tc={2000,5000};
call nlplm(rc,xres,"plgrad",x0,optn,con2,tc);
f = plgrad(xres); s = ssq(f);
if (s < 1.e-6) then xlb[ipar] = xres[ipar];
else xlb[ipar] = -1.0;

 end;


ci=xlb||xub;
create Profile from ci[colname={'Lower','Upper'} ] ;
append from ci;
close Profile;
run;
quit;


data WldCI; set WldCI;
Type= "Wald Method";
n=_n_;
if n<3 then delete;
keep Type Lower Upper;
run;

data profile; set profile;
Type= "Weighted Profile";
n=_n_;
if n<3 then delete;
keep Type lower upper;
run;


data WaldProf;
length Type $20.;
set WldCI profile;
```

```
run;

proc print data=WaldProf noobs;
title "Confidence interval of the difference of two paired binomial proportions in presence
 of missing data";
run;

%mend;
```

Illustration: Chang (2011) confidence interval

Consider the neurological data introduced in section 6.6.1 from Choi and Stablein [24] on 33 young meningitis participants. As stated before, in this study, $n_{00} = 6$, $n_{01} = 3$,$n_{10} = 8$,$n_{11} = 8$,$n = 25$,$w_1 = 4$, $l_1 = 6$, $w_2 = 2$, and $l_2 = 2$. All missing values of the data are assumed to be due to administrative reasons, and hence can be assumed to be missing at random (MAR). The following SAS code can be run to compute the 95% CI of the difference of proportions.

```
/*CI of the difference of proportions of the paired binomial with incomplete pairs*/
/*w1=u l1=m1 w2=v l2=v                                                            */
%multinomPF (n00=6 , n01= 3,n10=8 ,n11=8 , u=4 ,m1=6 ,v=2 ,m2=2, adj=0.5,adju=1, adjv=1,
 alpha=0.05);
```

The SAS output:

```
Confidence interval of the difference of two paired binomial proportions in presence of
 missing data

           Type                Lower       Upper

      Wald Method           -0.073154     0.37180
      Weighted Profile      -0.080473     0.36683
```

The SAS output shows 95% **TWC** and **TPFC** confidence intervals, respectively, as $(-0.073154, 0.37180)$ and $(-0.080473, 0.36683)$. In the SAS code the syntax *adj=0.5,adju=1, adjv=1* produce the corrected Tang-Wald and the corrected profile likelihood confidence intervals. In the SAS code, the syntax *adj=0,adju=0, adjv=0* would produce uncorrected Tang-Wald and the uncorrected profile likelihood confidence intervals. Note that, both **TWC** and **TPFC** intervals include zero, implying that the null hypothesis cannot be rejected at 5% level of significance. Hence, the neurological complication rates before and after treatment are plausibly the same.

6.7 Discussion and recommendation

In this chapter, we have discussed different methods of finding confidence interval for the difference of binomial proportions in paired binomial set up. The methods discussed are asymptotic, exact and Bayesian. For implementation of these intervals to real-life data sets SAS and R codes are provided. The applications of these intervals are also illustrated with real-life data sets. The illustrative examples clearly bring into the fore an important fact that every practitioner encounters, confidence intervals using different methods may lead to different inferences especially when the sample size is small. Therefore, it is often asked in practice when to use what?

Newcombe [62] and Fagerland et al. [33] compared the performances of the confidence intervals from different methods in paired binomial setup. The comparisons of those methods in either of these papers were somewhat limited. The recent paper by Fagerland et al. [33] studied coverage, expected lengths, and locations (using Mesial and Distal as suggested by Newcombe) by fixing one of the two parameters and varying the other in the p_1-space and p_2-space. Since $\Delta = p_1 - p_2$ involves two parameters in the parameter space $[0,1] \times [0,1]$, one needs to check performances of these methods in the entire parameter space. The following section presents the results of a simulation for $n = 10$ with odds ratio $\phi(= p_{11}p_{22}/p_{10}p_{01}) = 1$. Since $0 \leq p_1, p_2 \leq 1$, an uniform partition of 100 points are created in p_1 direction, and similarly, an uniform partition of 100 points are created in p_2 direction, i.e., a total of $100 \times 100 = 100^2$ (p_1, p_2) points are created in an unit square. Therefore, each grid point of this unit square represents a true Δ with $\Delta = p_1 - p_2$. For each of these Δs with points (p_1, p_2), the expected coverage probability (ECP) and expected length (ECW) are computed using the formula.

$$\text{ECPs} = \sum_{n_{11}=0}^{n} \sum_{n_{10}=0}^{n-n_{11}} \sum_{n_{01}=0}^{n-n_{11}-n_{10}} I(L \leq \Delta \leq U) f(n, n_{11}, n_{10}, n_{10})$$

and

$$\text{ECWs} = \sum_{n_{11}=0}^{n} \sum_{n_{10}=0}^{n-n_{11}} \sum_{n_{01}=0}^{n-n_{11}-n_{10}} (U - L) f(n, n_{11}, n_{10}, n_{10}),$$

where $I(.)$ is the indicator function, $[L, U]$ is confidence interval for any of the six procedures considered here, and

$$\begin{aligned} f(n, n_{11}, n_{10}, n_{01}) &= \frac{n!}{n_{11}!n_{10}!n_{01}!(n - n_{11} - n_{10} - n_{01})!} p_{11}^{n_{11}} p_{10}^{n_{10}} p_{01}^{n_{01}} \\ &\quad \times (1 - p_{1+} - p_{01})^{n-n_{11}-n_{10}-n_{01}} \end{aligned}$$

Finally, the distribution of the coverage and expected lengths of all of these 100^2 grid points are plotted using BliP plots (Lee and Tu [53]). Figure 6.9 presents the coverage probability distributions of all 100^2 grid points, and similarly Figure 6.10 presents the expected length distribution of all 100^2 grid points of the unit square. The vertical bars in each plotting symbol show the deciles of achieved coverage and the expected length distributions. A large

FIGURE 6.9
Distribution of the coverage probabilities of 95% confidence intervals for $n = 10$ for paired binomial proportions. AM=Agresti and Min MVR-WI=MOVER–Wilson, MVR-AC=MOVER–Agresti, MVR-JF=MOVER–Jeffreys, TG=Tango, PF=Weighted profile likelihood, COP= Copula-based weighted profile likelihood, and EX=unconditional exact method.

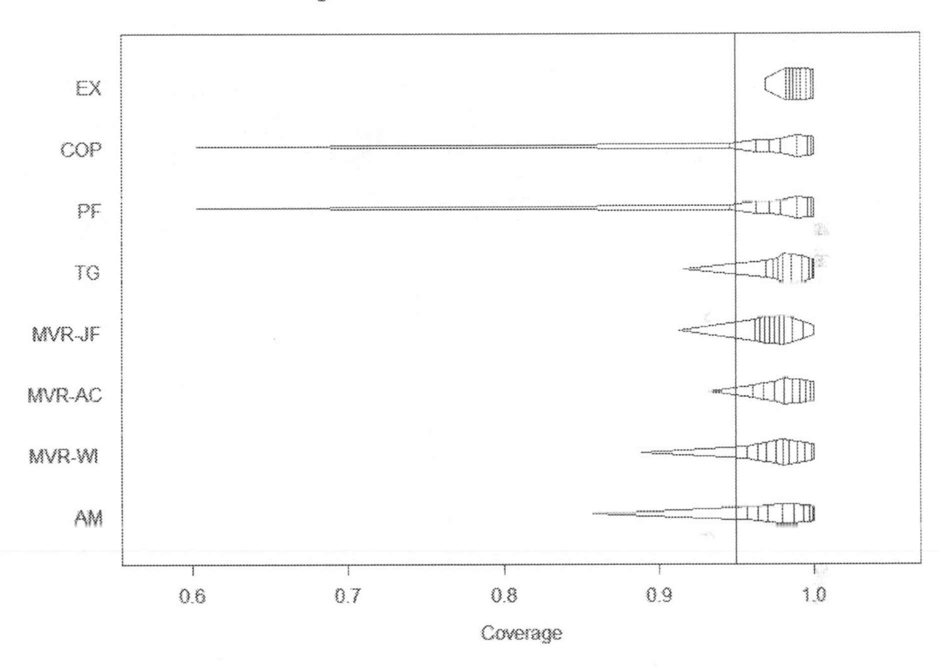

vertical line in Figure 6.9 is for the 95% level.

In Figure 6.9, the plot corresponding to unconditional exact method is aligned on the right side of the 95% level, confirming all of the (p_1, p_2) points of the total of $100 \times 100 = 100^2$ (p_1, p_2) points achieve nominal coverage levels. As stated in Section 6.5, for all the cases when both the off-diagonal elements are zero, confidence intervals are computed using the asymptotic score method as proposed by Tango. The long spikes corresponding to the methods COP and PF show undercoverage probabilities for less than 10% of the (p_1, p_2) points (as there is only one vertical bar to the left of the 95% level). The lowest coverage found in both of these methods is approximately 60%. The remaining asymptotic methods are approximately similar to each other in terms of achieving nominal coverage probabilities. As none of those plots

FIGURE 6.10
Distribution of the expected lengths of 95% confidence intervals of $n = 10$ for paired binomial proportions. AM=Agresti and Min MVR-WI=MOVER–Wilson, MVR-AC=MOVER–Agresti, MVR-JF=MOVER–Jeffreys, TG=Tango, PF=Weighted profile likelihood, COP= Copula-based weighted profile likelihood and EX=unconditional exact method.

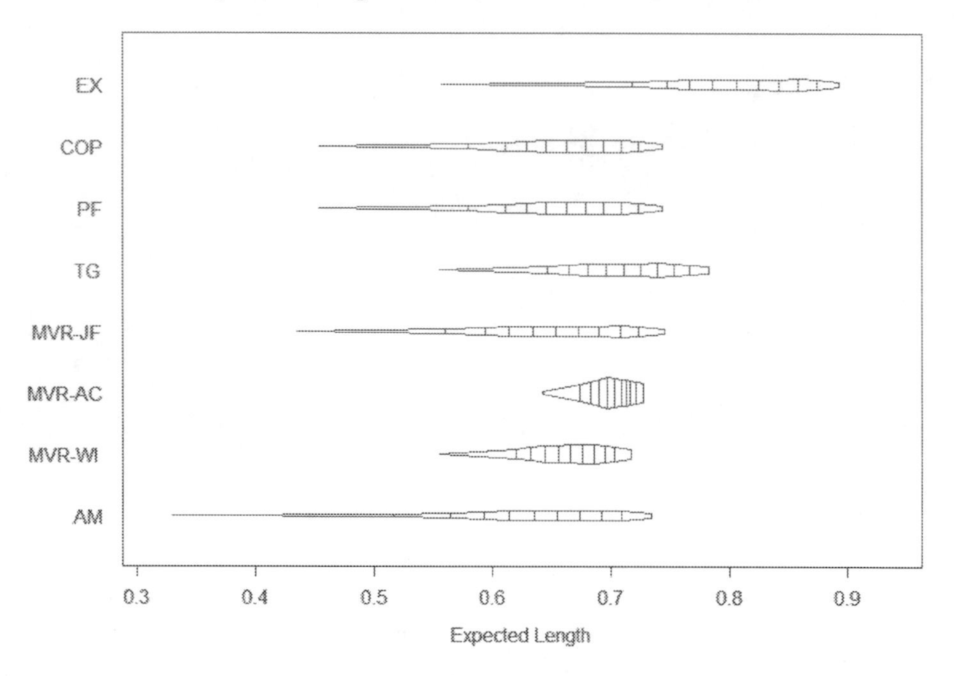

has any vertical bar to the left of the 95% level, less than 10% (p_1, p_2) points failed to attain the nominal 95% coverage.

Figure 6.10 shows the expected lengths of different methods. The plot corresponding to the exact method shows higher expected lengths than the other methods. Agresti and Min method appear to be the best followed by MOVER–Jeffreys method, weighted profile likelihood method, copula-based method, score method, MOVER–Wilson, and MOVER–Agresti and Coull method.

In a small sample set-up, a confidence interval using asymptotic methods may not attain the nominal coverage level. In those situations, even though the exact method is conservative, it is recommended that the exact method be used. When the sample size is moderate, any of the asymptotic methods can be used, as they are reasonably good in terms of achieving nominal coverage level, and CIs are less conservative than the exact method.

Section 6.6 includes a method of computing CI in the presence of missing data. These methods work well when the missing data mechanism is missing at random (MAR). For further details on CI with missing data, we refer to Tang et al.[90] and Pradhan et al. [71].

7

One Sample Count Data: Confidence Interval for Rate

7.1 Introduction

In this chapter, we will focus on the confidence interval methods for count data. Starting from the basic Poisson model, we will consider extensions to under and over-dispersed models and models with excess zero counts. The models we consider here are some of the most commonly used in this context but are not exhaustive. For Poisson distribution, the confidence interval of the mean is well-studied in the literature, but for others, the available methods are few and far between. This chapter includes interval estimation methods that we think would be most helpful to practitioners. We illustrate the methods with applications related to clinical research.

7.2 Poisson distribution

Suppose $X(t)$ represents the count of an event occurring in time in an interval of length t, and let the following assumptions hold:

- The chance of occurrence of an event in a small interval of time Δt is approximately equal to $\lambda\Delta t$
- The chance of occurrence of two or more events in a small interval of time Δt is negligible
- Events occurring over disjoint intervals of time are independent

Then, under these assumptions, the count $X(t)$ follows a Poisson distribution. Its probability mass function is given by

$$P_r(t) = P(X(t) = r) = \frac{e^{-\lambda t}(\lambda t)^r}{r!}, r = 0, 1, \ldots \infty \qquad (7.1)$$

where $\lambda(> 0)$ is the rate of occurrence of the event over time. We denote the

DOI: 10.1201/9781315169859-7

count in a unit interval $X(1)$ by X. Note that in some applications, instead of the count of an event occurring over time, $X(s)$ may represent the count over a space of size s. In that case, λ denotes the rate of occurrence of the event over space.

7.3 Confidence interval of rate parameter λ

Given an observed value x of X, a two-sided confidence interval of λ is denoted by $[\lambda_L(x), \lambda_U(x)]$. For brevity, the interval is often expressed as $[\lambda_L, \lambda_U]$. The confidence limits of a two-sided confidence interval satisfy the conditions $\lambda_L \geq 0$, $\lambda_U < \infty$. One-sided confidence intervals are given by $[0, \lambda_U]$ and $[\lambda_L, \infty)$. The confidence coefficient of a confidence interval $[\lambda_L, \lambda_U]$ is given by

$$\inf_{\lambda} C_{[\lambda_L,\lambda_U]}(\lambda) = \inf_{\lambda} P\left(\lambda_L(X) \leq \lambda \leq \lambda_U(X)\right) \tag{7.2}$$

where

$$\begin{aligned} P\left(\lambda_L(X) \leq \lambda \leq \lambda_U(X)\right) &= \sum_{x=0}^{\infty} I_{[\lambda_L(x),\lambda_U(x)]}(\lambda) P(X = x) \\ &= \sum_{x=0}^{\infty} I_{[\lambda_L(x),\lambda_U(x)]}(\lambda) \frac{e^{-\lambda}(\lambda)^x}{x!} \end{aligned} \tag{7.3}$$

and $I_{[a,b]}(\lambda) = 1$, if $a \leq \lambda \leq b$, and 0 otherwise. The confidence coefficient of the interval $[\lambda_L, \lambda_U]$ is said to be $1 - \alpha$ if

$$\inf_{\lambda} C_{[\lambda_L,\lambda_U]}(\lambda) \geq 1 - \alpha \tag{7.4}$$

i.e., $C_{[\lambda_L,\lambda_U]} \geq 1 - \alpha$ for all values of λ. The expected length of an interval for a given value of λ denoted by $L_{[\lambda_L,\lambda_U]}(\lambda)$ is given by

$$E\left(\lambda_U(X) - \lambda_L(X)\right) = \sum_{x=0}^{\infty} \left(\lambda_U(x) - \lambda_L(x)\right) P(X = x) \tag{7.5}$$

Often, independent counts of an event $X(t_1), X(t_2), ..., X(t_n)$ occurring at a rate λ are observed in time intervals of lengths, say, $t_1, t_2, ..., t_n$, respectively. Then $Y(t) = X(t_1) + X(t_2) + ... + X(t_n)$ has a Poisson distribution with parameter λt, where $t = t_1 + t_2 + ... + t_n$. Therefore, to find a confidence interval of λ from $Y(t)$, first, a confidence interval of λt is found based on $Y(t)$, and then its limits are simply divided by t.

7.3.1 Exact intervals

In this section, we will discuss confidence intervals derived from the exact probabilities based on Poisson distribution.

7.3.1.1 Garwood interval

This interval is proposed by Garwood [40]. It is similar to Clopper–Pearson's [26] interval for a single binomial proportion, and the method of construction guarantees a coverage of at least $1 - \alpha$. Given $X = x$, the confidence limits $\lambda_L(x)$ and $\lambda_U(x)$ are the largest and the smallest values of λ satisfying

$$P(X \geq x|\lambda_L) = \sum_{r=x}^{\infty} \frac{e^{-\lambda_L}(\lambda_L)^r}{r!} \leq \alpha/2 \tag{7.6}$$

and

$$P(X \leq x|\lambda_U) = \sum_{r=0}^{x} \frac{e^{-\lambda_U}(\lambda_U)^r}{r!} \leq \alpha/2 \tag{7.7}$$

respectively. Note that (Hastings and Peacock [47], p. 112)

$$P(X \leq x|\lambda) = P\chi^2_{2(x+1)} \geq 2\lambda). \tag{7.8}$$

Thus, the equations (7.6) and (7.7) can equivalently be written as

$$P(\chi^2_{2x} > 2\lambda_L(x)) = 1 - \frac{\alpha}{2} \tag{7.9}$$

and

$$P(\chi^2_{2(x+1)} > 2\lambda_U(x)) = \frac{\alpha}{2}, \tag{7.10}$$

respectively. Therefore, we have $\lambda_L(x) = \frac{1}{2}\chi^2_{2x,\alpha/2}$ and $\lambda_U(x) = \frac{1}{2}\chi^2_{2(x+1),1-\alpha/2}$. These limits are given by Garwood [40], Przyborowski and Wilenski [75] and also discussed by Hald [46](pp. 722–723), Liddel [55] and Ulm [93]. Almost all statistical software includes the method as one of the exact procedures of finding the confidence interval of λ. For one-sided intervals, λ_L and $\lambda_U(< \infty)$ are obtained by solving the above inequalities replacing $\alpha/2$ with α.

7.3.1.2 Blaker's interval

We have discussed Blaker's method for finding the exact interval for binomial proportion π in Section 3.5.4 of chapter 3. The method is applicable here, except that the distribution of X is now Poisson with parameter λ. Like Clopper-Pearson intervals, Garwood intervals are often too wide and have coverage probability considerably higher than $1-\alpha$. Blaker's intervals, instead, provide coverage probabilities closer to the nominal level, especially for λ near 0. It is clearly an improvement over the Garwood interval.

7.3.1.3 Mid-P interval

Kulkarni et al. [51] consider the interval of λ obtained by inverting the acceptance region of an equal tailed exact size α randomized test for $\lambda = \lambda_0$

against all alternatives. If γ, the randomization probability at the observed value x of X, is set to 1, the method reduces to Garwood interval. Clearly, such intervals would be conservative. On the other hand, intervals obtained by taking γ equal to zero may be liberal. As a compromise, an interval called the Mid-P interval is obtained by taking γ equal to 0.5. Given the observed value x of X, the lower and upper confidence limits of the Mid-P interval are obtained by solving the equations

$$P(\chi^2_{2(x+1)} \leq 2\lambda_L(x)) + P(\chi^2_{2x} \leq 2\lambda_L(x)) = \alpha$$
$$P(\chi^2_{2(x+1)} \leq 2\lambda_U(x)) + P(\chi^2_{2x} \leq 2\lambda_U(x)) = 2 - \alpha \quad (7.11)$$

An approximate solution to the above equations is also provided by Kulkarni et al. [51].

Illustration: Exact intervals

To illustrate the computation of the exact confidence intervals discussed above, we consider the data ([59]) from a clinical study of osteoporotic patients with high-risk vertebral fractures. The patients were randomized to either a treatment or a placebo. For illustration, we consider only the data from the treatment arm. The data are given in Table 7.1.

TABLE 7.1
Vertebral fracture data

subjid	x	prsnyr	subjid	x	prsnyr	subjid	x	prsnyr
1	1	2.9897	11	3	2.9596	21	1	1.2676
2	2	2.9843	12	1	2.9843	22	2	2.9541
3	0	2.9897	13	0	2.9925	23	0	2.3874
4	1	2.9843	14	1	1.4346	24	0	1.2868
5	2	2.9459	15	1	3.0116	25	0	1.2813
6	0	3.0445	16	0	2.9733	26	2	2.7625
7	1	3.0582	17	2	3.0637	27	2	2.4504
8	0	2.9897	18	1	2.9706	28	4	2.8747
9	1	1.2567	19	0	1.1882			
10	0	2.9678	20	0	2.9733			

In Table 7.1, for each subject, x is the count of vertebral fractures during the period *prsnyr* (person year). Here, the duration is called an offset, and it is used to normalize the mean. A point estimate and 95% exact confidence intervals of the rate parameter λ can be computed using R package *exactci* as shown below.

R code: Garwood interval

```
Input <- read.csv('Fractures.csv')
library(exactci)

y<-Input$y
ptr<-Input$prsnyr
#--------getting the total count y and the total duration----
x<-sum(y)
py<-sum(ptr)

#-------CI based on exact method-------------------------------
poisson.exact(x,py,tsmethod="central",conf.level=0.95)


> poisson.exact(x,py,tsmethod="central",conf.level=0.95)

Exact two-sided Poisson test (central method)

data:  x time base: py
number of events = 28, time base = 72.027, p-value = 5.672e-09
alternative hypothesis: true event rate is not equal to 1
95 percent confidence interval:
 0.2583159 0.5618397
sample estimates:
event rate
 0.3887415
```

Note that the point estimate of λ is 0.3887415, and the 95% Garwood interval is $[0.2583159, 0.5618397]$.

R Code: Mid-p and Blaker's intervals

```
#-------CI based on mid-p correction---------------------------
poisson.exact(x,py,tsmethod="central",midp=TRUE, conf.level=0.95)
#-------CI based on Blaker---------------------------------------
poisson.exact(x,py,tsmethod="blaker",conf.level=0.95)


> poisson.exact(x,py,tsmethod="central",midp=TRUE, conf.level=0.95)

Exact two-sided Poisson test (central method), mid-p version

data:  x time base: py
number of events = 28, time base = 72.027, p-value = 3.915e-09
alternative hypothesis: true event rate is not equal to 1
95 percent confidence interval:
 0.2634017 0.5543150
sample estimates:
event rate
 0.3887415

> poisson.exact(x,py,tsmethod="blaker",conf.level=0.95)

Exact two-sided Poisson test (Blaker's method)

data:  x time base: py
number of events = 28, time base = 72.027, p-value = 4.583e-09
alternative hypothesis: true event rate is not equal to 1
95 percent confidence interval:
 0.2644900 0.5562308
sample estimates:
event rate
 0.3887415
```

The 95% Mid-p interval and Blaker's interval of λ are [0.2634017 0.5543150] and [0.2644900 0.5562308], respectively. The Blaker's interval is more precise (shorter expected length) than the Garwood interval with guaranteed nominal coverage.

7.3.2 Asymptotic intervals

7.3.2.1 Wald-interval

For sufficiently large values of λ (more than 100 (Armitage et al., 2002)), given the observed value x of X, the confidence limits of the Wald-interval with confidence coefficient $1 - \alpha$ are given by

$$\begin{aligned} \lambda_L(x) &= x - z_{\alpha/2}\sqrt{x} \\ \lambda_U(x) &= x + z_{\alpha/2}\sqrt{x}. \end{aligned} \tag{7.12}$$

The interval is based on the result that the distribution of $(X - \lambda)/\sqrt{X}$ is approximately a standard normal for sufficiently large values of λ. Often a continuity corrected version of Wald-interval is used with

$$\begin{aligned} \lambda_L(x) &= x - 0.5 - z_{\alpha/2}\sqrt{x} \\ \lambda_U(x) &= x + 0.5 + z_{\alpha/2}\sqrt{x}. \end{aligned} \tag{7.13}$$

The continuity correction may be ignored if x is large.

7.3.2.2 Score-interval

The confidence limits of the score interval are obtained by solving the equations $(x - \lambda)/\sqrt{\lambda} = \pm z_{\alpha/2}$ for λ. The lower and upper limits are

$$\begin{aligned} \lambda_L(x) &= x + 0.5z^2_{\alpha/2} - z_{\alpha/2}\sqrt{x + 0.25z^2_{\alpha/2}} \\ \lambda_U(x) &= x + 0.5z^2_{\alpha/2} + z_{\alpha/2}\sqrt{x + 0.25z^2_{\alpha/2}}. \end{aligned} \tag{7.14}$$

The confidence limits of the score interval with continuity correction are given by

$$\begin{aligned} \lambda_L(x) &= x - 0.5 + 0.5z^2_{\alpha/2} - z_{\alpha/2}\sqrt{x - 0.5 + 0.25z^2_{\alpha/2}} \\ \lambda_U(x) &= x + 0.5 + 0.5z^2_{\alpha/2} + z_{\alpha/2}\sqrt{x + 0.5 + 0.25z^2_{\alpha/2}}. \end{aligned} \tag{7.15}$$

Replacing $z_{\alpha/2}$ in (7.15) by the upper $100(\alpha/2)$ percentile point of a distribution that provides a better approximation to the distribution of $(x - \lambda)/\sqrt{\lambda}$ than $N(0, 1)$ (Molenaar, 1970), Sahai and Khurshid [79] proposed a confidence interval of λ with the limits given by

$$\begin{aligned}\lambda_L(x) &= x + (2z^2_{\alpha/2} + 1)/6 - \left[0.5 + \sqrt{z^2_{\alpha/2}\left(x - 0.5 + \frac{z^2_{\alpha/2} + 2}{18}\right)}\right] \\ \lambda_U(x) &= x + (2z^2_{\alpha/2} + 1)/6 + \left[0.5 + \sqrt{z^2_{\alpha/2}\left(x + 0.5 + \frac{z^2_{\alpha/2} + 2}{18}\right)}\right].\end{aligned} \tag{7.16}$$

.

Illustration: Asymptotic intervals

We illustrate the computation of the above asymptotic intervals using the vertebral fracture data (cf. Table 8.1) described above. The SAS macro is given below:

SAS Code

```
%macro asypoi (x= , conflev=0.95, Py= , type='wld');

data _CI;
x=&x;
z = abs(probit((1-&conflev)/2));
%if &type='wld' %then %do;
          lower=x-z*sqrt(x);
          upper=x+z*sqrt(x);
          LowerCI=lower/&Py;
          UpperCI=upper/&Py;
%end;
%else %if &type='wldcc' %then %do;
          lower=x-z*sqrt(x)-0.5;
          upper=x+z*sqrt(x)+0.5;
          LowerCI=lower/&Py;
          UpperCI=upper/&Py;
%end;
%else %if &type='sc' %then %do;
          lower=x+0.5*z**2 -z*sqrt(x+0.25*z**2);
          upper=x+0.5*z**2 +z*sqrt(x+0.25*z**2);
          LowerCI=lower/&Py;
          UpperCI=upper/&Py;
%end;
%else %if &type='scc' %then %do;
          lower=x+0.5*z**2 -z*sqrt(x+0.25*z**2 -0.5)-0.5;
          upper=x+0.5*z**2 +z*sqrt(x+0.25*z**2 +0.5)+0.5;
          LowerCI=lower/&Py;
          UpperCI=upper/&Py;
%end;

%else %if &type='ssk' %then %do;
          lower=x+(2*z**2 +1)/6-(0.5+sqrt(z**2 * (x-0.5+(z**2 +2)/18)));
          upper=x+(2*z**2 +1)/6+(0.5+sqrt(z**2 * (x+0.5+(z**2 +2)/18)));
          LowerCI=lower/&Py;
```

```
          UpperCI=upper/&Py;
%end;

run;
proc print data=_ci noobs;
title "%sysevalf(&conflev*100) percent confidence interval of the Poisson rate using
 method=&type    ";
var LowerCI UpperCI;
run;

%mend;
```

The SAS code for the asymptotic intervals is given above. We consider the computation of the Wald interval of vertebral fracture rate and then indicate the changes to be made for other intervals. From Table 8.1, we observe that in total, 28(x) vertebral fracture events occurred in a total of 72.027 person-year duration.

SAS Code: Wald interval

```
%asypoi (x=28, conflev=0.95, Py=72.027 , type='wld');
                   LowerCI    UpperCI

                   0.24475     0.53273
```

In the first line of the SAS macro (%asypoi), we plug in the values of the count (x) and the total person year (*Py=*). Notice that *conflev=*0.95 is for 95% confidence interval, and *type='wld'* is for Wald method.

For other intervals, one needs to change *type='wld'* to the corresponding type, like *type='wldcc'* for continuity corrected Wald interval, *type='sc'* for score interval, *type='scc'* for continuity corrected score interval, and *type='ssk'* for the Sahai and Khurshid interval. In Table 7.2, we report all the 95% confidence intervals of vertebral fracture that we discuss here.

7.3.2.3 The likelihood ratio interval

The likelihood ratio interval is an asymptotic interval, which deserves special attention because of its superiority over others as observed by Brown et al. [16]. It is given by $\{\lambda| - 2ln\Lambda \leq \chi^2_{\alpha,1}\}$, where $\Lambda = L(\lambda)/Sup_\lambda L(\lambda)$ and $L(\lambda) = e^{-\lambda}\lambda^x/x!$ is the likelihood function. The coverage and expected length properties of this interval are discussed in detail in Brown et al. [16].
Among different asymptotic intervals, Brown et al. [16] find the likelihood ratio interval to be the best, closely followed by the Jeffreys interval and the score interval. The Wald interval performs poorly compared to the other three.

SAS code: Likelihood ratio interval

Using SAS's PROC GENMOD, the confidence interval based on likelihood ratio can be computed using the following SAS code:

```
data fractures;
input n @@;
do subjid=1 to n;
    input y prsnyr @@;
    output;
end;
drop n;
datalines;
28 1 2.9897 2 2.9843 0 2.9897 1 2.9843 2 2.9459 0 3.0445 1 3.0582
0 2.9897 1 1.2567 0 2.9678 3 2.9596 1 2.9843 0 2.9925 1 1.4346
1 3.0116 0 2.9733 2 3.0637 1 2.9706 0 1.1882 0 2.9733 1 1.2676
2 2.9541 0 2.3874 0 1.2868 0 1.2813 2 2.7625 2 2.4504 4 2.8747
;
run;

data _one; set fractures;
    logt=log(prsnyr);
run;

ods output ParameterEstimates=_out; /*getting the ods output as SAS data _out */
proc genmod data=_one;
    model y= /offset=logt dist=p link=log lrci;
run;

/*-----------------exponentiating the estimates--------------------------------*/
data _out1; set _out;
    Rate=exp(Estimate);
    Lower=exp(LowerLrCL);
    Upper=exp(UpperlrCL);
    if Parameter="Intercept";
    Keep Rate Lower Upper;
run;

proc print data=_out1 noobs;
title "Likelihood ratio test based confidence interval";
run;
```

Notice that the log transformation of the offset has been done in the data step. In the PROC GENMOD code, *dist=p link=log lrci* invokes the Poisson distribution to compute the mean Poisson rate using log link (link=log). Note that, in this model, the intercept is the logarithm of the mean rate parameter of interest. The option *lrci* invokes the confidence interval of the rate parameter inverting likelihood ratio test using the specified offset. The option offset has been specified in the option *offset=logt*. Since we have specified the option *link=log*, the estimate is computed in the log scale; hence the confidence interval for the rate parameter can be obtained by exponentiating the resulting confidence limits. The final SAS output is shown below:

```
        Likelihood ratio test based confidence interval

                  Rate        Lower       Upper

                0.38874     0.26195     0.55103
```

The mean rate of the fracture count is shown as 0.3887 and the corresponding confidence interval due to asymptotic likelihood ratio test is $(0.26195, 0.55103)$.

7.3.3 Bayesian interval

7.3.3.1 The Jeffreys' interval

This is the equal-tailed Bayesian posterior predictive interval of λ using Jeffreys prior, which is in this case proportional to $\lambda^{-1/2}$. The posterior distribution of λ is then Gamma $(x+1/2)$. Thus, $100(1-\alpha)\%$ equal tailed Jeffreys interval for λ is given by $[G_{1-\alpha/2}(x+1/2), G_{\alpha/2}(x+1/2)]$, where $G_\alpha(x)$ is the 100α upper percentile point of Gamma(x) distribution with the probability density function given by $f(u) = 1/\Gamma(x)e^{-u}u^{x-1}, u > 0$.

R code: The Jeffreys' interval

The Jeffreys interval can be easily implemented in SAS by writing a SAS macro. The R software package *ratesci* has implemented this method along with some other methods stated in this chapter. For the vertebrate fracture data, the following R script can be run:

```
library(ratesci)
Input <- read.csv('Fractures.csv')
y<-Input$y
ptr<-Input$prsnyr
x<-sum(y)
py<-sum(ptr)
jeffreysci(x, n=py, ai = 0.5,  cc = 0, level = 0.95, distrib = "poi")
```

R software shows the following output:

```
         Lower     Upper       est
[1,] 0.2639745 0.5536248 0.3910652
```

7.3.4 Remarks on the exact, asymptotic and Bayesian intervals

In Table 7.2, we summarize the 95% confidence intervals of the mean rate of the fracture count using all the methods that we have described here.

Among the exact intervals, the Garwood and the Blaker's intervals are guaranteed to attain the nominal coverage level. However, it is not the case with the mid-p-corrected interval and consequently is less conservative. Therefore, the width of the interval is a little shorter than the other exact methods. Among the asymptotic methods, the width of the likelihood ratio interval is the shortest among all, barring Wald's interval. As stated in section 7.3.2.3, Wald's interval performs poorly in most situations, and in fact, Brown et al. ([16]) recommend banning its use in practice. The length of Jeffreys' interval is also very close to that of the likelihood ratio interval, and Brown et al. ([16]) recommend it as a good interval. It should be noted, however, that both the likelihood ratio and Jeffreys' interval are a little shorter than Blaker's interval, but these intervals may fail to preserve the nominal coverage probability.

TABLE 7.2
95% confidence intervals of the mean rate of vertebrate fracture count

Method	Lower	Upper	Length
Wald	0.24475	0.53273	0.28798
Wald with CC	0.23781	0.53967	0.30186
Score	0.26897	0.56185	0.29288
Score with CC	0.2633	0.57005	0.30675
Sahai and Khurshid	0.25836	0.56187	0.30351
Likelihood ratio	0.26195	0.55103	0.28908
Jeffreys'	0.26397	0.55362	0.28965
Garwood	0.258316	0.56184	0.303524
Mid-p-corrected	0.263402	0.554315	0.290913
Blaker	0.26449	0.556231	0.291741

7.4 Confidence interval for mean: Other count data models

We now discuss confidence intervals for the mean of a few other commonly used models for count data. The models discussed here are particularly useful for the data showing either extra variation (variance significantly more than the mean) or excess zero counts (zero frequency is significantly more than what is expected in a count model), or both. For these models, a commonly used confidence interval for the mean (μ) with confidence coefficient $1-\alpha$ is the Wald interval, which is given by $\widehat{\mu} \pm z_{\alpha/2} \times \widehat{se}$, where $\widehat{\mu}$ is an estimate of μ, and $\widehat{se}$ is an estimate of its standard error. This interval is based on the assumption that the distribution of $(\widehat{\mu}-\mu)/\widehat{se}$ is approximately $N(0,1)$ for moderate to large n. In the following discussions, we illustrate the computation of the confidence intervals based on ML estimates as well as confidence intervals based on the sample mean and the sample variance. For small sample sizes, bootstrap confidence intervals could sometimes be useful.

7.4.1 Negative binomial distribution

In drug development studies, subjects experiencing certain types of recurrent events are quite common. For example, in clinical trials, often the subjects experience recurrent adverse events such as infection, epileptic seizure and hospitalization. The rates of occurrences of such events may vary among the subjects. Since the analysis using Poisson distribution does not address the heterogeneity in rates among the subjects, a correction for the extra variation is needed.

Greenwood and Yule [45] introduce Negative Binomial distribution (**NBD**) as a model for extra-Poisson variation. Assuming a gamma distribution for the

Poisson mean (λ) with $E(\lambda) = \mu$ and variance $var(\lambda) = \mu^2 d$, the Negative Binomial distribution is obtained as the marginal distribution of Y with probability density function

$$f(y|\mu, d) = \frac{\Gamma(y + d^{-1})}{y!\Gamma(d^{-1})} \left(\frac{d\mu}{1 + d\mu}\right)^y \left(\frac{1}{1 + d\mu}\right)^{d^{-1}} \tag{7.17}$$

for $y = 0, 1, 2, \ldots; \mu > 0$ and $d > 0$. The parameter d is known as the over-dispersion parameter (see Lawless [52], Cameron and Trivedi [18]). Note that the limiting form of this distribution is Poisson as d goes to zero. The negative Binomial model is extensively used for modelling count data showing extra Poisson variation. For **NBD**, the mean is μ, and the variance $\mu(1 + d\mu)$ is greater than the variance of the Poisson distribution.

Let $Y_1, Y_2, ..., Y_n$ be a random sample from the **NBD**. There are several methods to estimate mean μ and the over-dispersion parameter d (see Saha and Paul [78], Robinson and Smythe [76]). For example, the Method of Moments (MM) estimates are $\widehat{\mu}_M = \bar{y}$ and $\widehat{d}_M = (s^2 - \bar{y})/\bar{y}^2$ where $\bar{y}$ and s^2 are the sample mean and sample variance of Y*'s*, respectively. The maximum likelihood estimates $\widehat{\mu}_L$ and $\widehat{d}_L$ are obtained by maximizing the log-likelihood function given by

$$l(\mu, d) = \sum_{i=0}^{n} [y_i log(\mu) - \left(y_i + \frac{1}{d}\right) log(1 + d\mu) + \sum_{j=0}^{(y_i - 1)} log(1 + dj)] \tag{7.18}$$

with respect to μ and d. Among other methods of estimation, the quasi-likelihood method by Wedderburn [98], and the pseudo-likelihood method by Carrol and Ruppert [19] may be used.

Illustration: Confidence intervals for negative binomial distribution

For illustration, we consider the spinal tumour count data (cf. Joe and Zhu [48]) for patients with neurofibromatosis 2 (NF2) diseases. Table 7.3 provides a quick statistical summary of the data. Note that the sample mean of tumor

TABLE 7.3
Summary of tumor counts with neurofibromatosis 2 (NF2) disease

N	Mean	Variance	Min	Max	Percentage of Zeros
158	4.335	62.403	0	50	0.443

counts ($\bar{y} = 4.335$) is substantially lower than the sample variance $s^2 = 62.403$. Clearly, the data show a strong overdispersion. Also, zero count occurs with a high frequency (44%). For modeling count data with such a high frequency of zero, zero-inflated distributions are usually used. We defer its discussion to the end of this chapter.

The histogram of the data (Figure 7.1) clearly shows that the data distribution is heavily right-skewed with a high frequency at zero. Clearly, Poisson is not the right model for this data; hence we fit **NBD**. SAS offers multiple PROCS for fitting **NBD** (for example, PROC GENMOD, COUNTREG, NLMIXED, GLIMMIX). Here, we use PROC GLIMMIX. The SAS code is given here.

FIGURE 7.1
Tumor distribution of 158 patients.

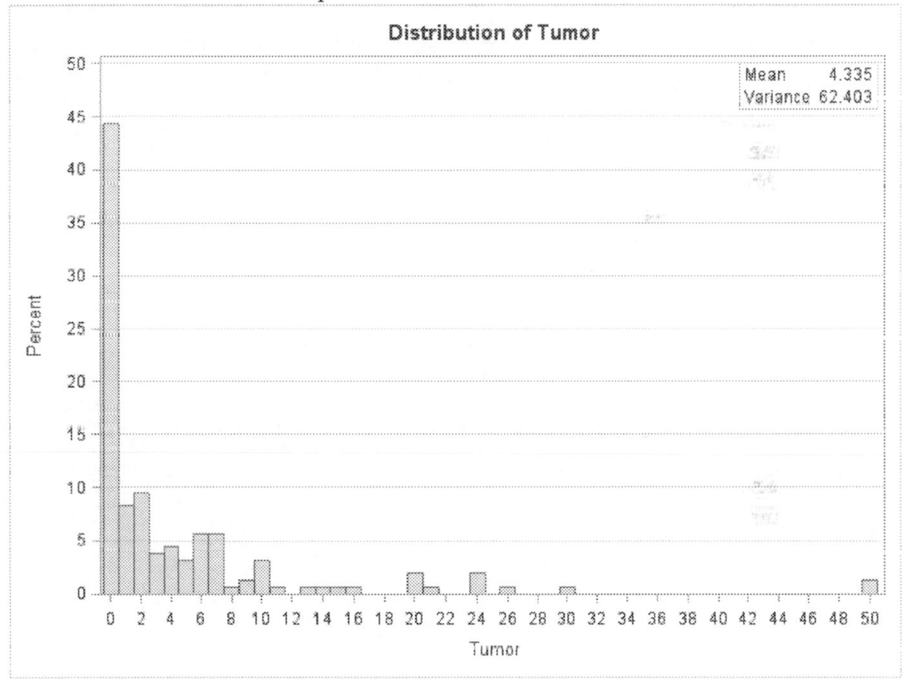

SAS Code

```
data Tumor;
input Tumor count @@;
cards;
0 70 1 13 2 15 3 6 4 7 5 5
6 9 7 9 8 1 9 2 10 5 11 1 13 1
14 1 15 1 16 1 20 3 21 1
24 3 26 1 30 1 50 2
;
run;
proc glimmix data=tumor ;
model tumor= /solution dist=NB link=identity /* nb=Negative Binomial, if link=log
                                               then for mu=exp(intercept)*/;
```

```
NLOPTIONS TECH=NRRIDG;
freq count;
run;
```

Following is the SAS output.

```
Convergence criterion (GCONV=1E-8) satisfied.

                 Fit Statistics

         -2 Log Likelihood              740.07
         AIC  (smaller is better)       744.07
         AICC (smaller is better)       744.15
         BIC  (smaller is better)       750.20
         CAIC (smaller is better)       752.20
         HQIC (smaller is better)       746.56
         Pearson Chi-Square             151.24
         Pearson Chi-Square / DF          0.96


               Parameter Estimates

                          Standard
Effect       Estimate       Error      DF    t Value    Pr > |t|

Intercept      4.3354      0.6403     157       6.77      <.0001
Scale          3.2159      0.4614       .          .           .
```

Since we fit a model with intercept only, in the above output, the Effect=Intercept is the maximum likelihood estimate of mean μ, which is also the sample mean. The Effect=Scale is the maximum likelihood estimate of the overdispersion parameter. Hence, for this data $\widehat{\mu} = 4.3354$, the estimate of its standard error is $\widehat{se} = 0.6403$; $\widehat{d} = 3.2159$, and the corresponding estimate of its standard error is 0.4614. Thus the estimate of the variance V can be computed as

$$\begin{aligned}\widehat{V} &= \widehat{\mu}(1 + \widehat{d}\widehat{\mu}) \\ &= 4.3354(1 + 3.2159 \times 4.3354) \\ &= 64.78.\end{aligned}$$

Note that the model-based estimate of V is 64.78, slightly higher than the sample variance 62.403. The confidence interval of μ with confidence coefficient 0.95 is given by

$$\begin{aligned}&\widehat{\mu} \pm z_{\alpha/2} \times \widehat{se} \\ &= 4.3354 \pm 1.96 \times 0.6403,\end{aligned}$$

which reduces to $(3.08, 5.59)$. On the other hand, if we use sample variance as an estimate of population variance, the estimate of the standard error of sample mean $\widehat{se}$ becomes $s/\sqrt{n} = 62.403/158 = 0.6284$, an estimate lower than the model-based estimate 0.6403. The corresponding confidence interval for μ reduces to $(3.10, 5.57)$, which is slightly shorter than the ML-based interval.

Assuming Poisson model for the same data, the Clopper–Pearson interval of μ is $(4.017, 4.673)$, and that of the Mid-p interval is $(4.02, 4.669)$. Clearly, these intervals are substantially shorter, as they fail to capture the additional variability (overdispersion) present in the data.

7.4.2 Generalized Poisson distribution

Consul and Jain [30] propose a distribution for count data, known as Generalized Poisson distribution (**GPD**), which could model both under and over dispersion, and its density function is given by

$$f(y|\lambda,\xi) = \frac{\lambda(\lambda+y\xi)^{y-1}e^{-(\lambda+y\xi)}}{y!} \tag{7.19}$$

$y = 0, 1, 2, \ldots$
for $\lambda > 0$, and $-1 < \xi < 1$ such that $f(y|\lambda,\xi) = 0$ for $y \geq m$ if $\lambda + m\xi \leq 0$.

Its mean and variance are given by $\frac{\lambda}{1-\xi}$ and $\frac{\lambda}{(1-\xi)^3}$, respectively. The model is over-dispersed for $\xi > 0$, and under-dispersed for $\xi < 0$. For $\xi = 0$, it reduces to a Poisson distribution. Like **NBD**, **GPD** is also a Poisson mixture (Joe and Zhu [48]). However, compared to **NBD** it is heavier tailed and has less mass at zero. For most of the over-dispersed count data, **NBD** and **GPD** give very close fit. But for modeling count data with a high frequency of zero, the zero-inflated versions of these distributions may differ significantly(Joe and Zhu [48]). We will consider this in the next subsection.

Reparameterizing the model in terms of the mean as μ, we obtain from the above $\lambda = \mu(1-\xi)$. Hence the kernel of the log-likelihood (7.19) in terms of μ and ξ is

$$\begin{aligned} l(\mu,\xi;y) &= log\{\mu(1-\xi)\} + (y-1)log\{\mu - \xi(\mu - y)\} \\ &\quad - \{\mu - \xi(\mu - y)\} - log\Gamma(y+1). \end{aligned} \tag{7.20}$$

The maximum likelihood estimates of μ and ξ, say, $\widehat{\mu}$ and $\widehat{\xi}$, and their respective estimated standard errors $\widehat{se}_{\mu}$ and $\widehat{se}_{\xi}$ could be obtained by using procedures such as PROC NLMIXED or PROC GLIMMIX in SAS/STAT. Thus the ML-based interval of μ with confidence coefficient $1-\alpha$ given by $\widehat{\mu} \pm z_{\alpha/2} \times \widehat{se}_{\mu}$. However, since the $\widehat{\mu}$ is also the sample mean $\bar{y}$, one can also estimate the standard error of $\widehat{\mu}$ by $s/\sqrt{n}$, and the resulting confidence interval is $\bar{y} \pm z_{\alpha/2} \times s/\sqrt{n}$

Illustration: Confidence intervals for generalized Poisson distribution

To illustrate the GPD model, we consider the tumor data introduced in section 7.4.1. For maximization of the log-likelihood (7.20), we need to convert the frequency distribution of the count data into the case data.

SAS code

```
data Tumor;
input Tumor count @@;
cards;
0 70 1 13 2 15 3 6 4 7 5 5
6 9 7 9 8 1 9 2 10 5 11 1 13 1
14 1 15 1 16 1 20 3 21 1
24 3 26 1 30 1 50 2
;
run;

/*Converting all frequency counts and creating a count variable y*/
data one; set tumor;
   do i=1 to count;
      y=tumor;
      output;
   end;
drop i;
run;
```

/*Optimisation of the log-likelihood equation (8.20)*/

```
proc glimmix data=one ;
   model y= / link=identity s;
   /*--Scale parameter needs to be converted back to _phi_----------*/
   xi = (1 - 1/exp(_phi_));
   if xi>=1 then xi=0.99;
   _variance_ = _mu_ / (1-xi)/(1-xi);
   if (_mu_=.) or (_linp_ = .) then _logl_ = .;
   else do;
      mustar = _mu_ - xi*(_mu_ - y);
      if ((mustar < 0) or (_mu_*(1-xi) < 0))
         then _logl_ = 1E-2;
      else do;
      /*----------log-likelihood equation-------------------------------*/
        _logl_ = log(_mu_*(1-xi)) + (y-1)*log(mustar) -mustar -lgamma(y+1);
      end;
   end;
run;
```

Following is the SAS output.

```
Convergence criterion (GCONV=1E-8) satisfied.

                  Fit Statistics

       -2 Log Likelihood            748.79
       AIC  (smaller is better)     752.79
       AICC (smaller is better)     752.87
       BIC  (smaller is better)     758.92
       CAIC (smaller is better)     760.92
       HQIC (smaller is better)     755.28
```

```
           Pearson Chi-Square                  100.16

                          Fit Statistics

           Pearson Chi-Square / DF               0.64


                        Parameter Estimates

                                 Standard
Effect         Estimate          Error         DF      t Value     Pr > |t|

Intercept        4.3354         0.7868        157         5.51       <.0001
Scale            1.5581         0.1730          .            .            .
```

Since in PROC GLIMMIX code *link=identity* option is used, the estimate of μ corresponds to the estimate of the intercept. In the SAS output $\widehat{\mu} = 4.3354$ and $\widehat{se}_{\mu} = 0.7868$ are shown. Note that $\widehat{\mu} - 4.3354$ is the same as the sample mean. In the PROC GLIMMIX code, for computational advantage, the parameter ξ is reparametrized as $\xi = (1 - 1/exp(\phi))$, where ϕ is the new parameter. In the GLIMMIX output, the estimate of the scale is, in fact, the estimate of the parameter ϕ. Hence, the estimate of the parameter ζ is,

$$\begin{aligned}\widehat{\xi} &= 1 - \frac{1}{exp(\widehat{\phi})} \\ &= 1 - \frac{1}{exp(1.5581)} \\ &= 0.7895\end{aligned}$$

Hence, the ML estimate of variance from the model is given by

$$\begin{aligned}V &= \frac{\widehat{\mu}}{(1-\widehat{\xi})^2} \\ &= \frac{4.3354}{(1-0.7895)^2} \\ &= 97.809\end{aligned}$$

Clearly, 97.809 is much higher than the estimated variance 64.78 from the **NBD** and the sample variance 62.403. Thus, the 95% interval of μ is given by

$$\begin{aligned}&\widehat{\mu} \pm z_{\alpha/2} \times \widehat{se}_{\mu} \\ &= 4.3354 \pm 1.96 \times 0.7868,\end{aligned}$$

which is $[2.793, 5.8775]$. It is wider than the ML-based confidence interval $[3.08, 5.59]$ obtained by using **NBD**, or the interval $[3.10, 5.57]$ based on the sample mean and the sample variance. In this connection, it should be noted from the values of different goodness of fit criteria that both **NBD** and **GPD** fit the data well.

7.4.3 Zero-inflated models

In real-life applications, excess zero count (more than what could be explained by a count model) is often observed in count data. All the models discussed so far are not appropriate to describe such data. A mixture of a point mass at zero and a suitable model for count data is used for modelling this type of data. Such models are known as zero-inflated models. Thus, in a zero-inflated model, zeros occur from two sources – a point mass at zero and a count model. Any count model can be extended to a zero-inflated model. In the following, we discuss some commonly used zero-inflated models.

7.4.3.1 Confidence intervals for zero-inflated Poisson distribution (ZIPD)

The distribution is a mixture of a point mass at 0 with probability p and a Poisson(λ) distribution with probability $(1-p)$. Hence, the distribution of the count Y can be written as

$$Pr(Y = y) = \begin{cases} p + (1-p)\exp(-\lambda) & y = 0 \\ (1-p)\exp(-\lambda)\lambda^y/y! & y > 0. \end{cases} \quad (7.21)$$

The mean and variance are given by $\mu = \lambda\pi$ and $V = \lambda\pi + \lambda^2\pi(1-\pi)$, respectively, where $\pi = 1 - p$. If $p = 0$, then the model reduces to Poisson with mean λ.

Illustration: Zero-inflated Poisson (ZIPD)

For illustrative purposes, we use the tumor data introduced above, which has excess (44%)zero count.

SAS Code

```
data Tumor;
input Tumor count @@;
cards;
0 70 1 13 2 15 3 6 4 7 5 5
6 9 7 9 8 1 9 2 10 5 11 1 13 1
14 1 15 1 16 1 20 3 21 1
24 3 26 1 30 1 50 2
;
run;

proc genmod data=tumor;
   model tumor= /dist=zip;
   freq count;
   zeromodel /link=logit;
/*output out=_o pzero=prbzero;*/
run;
```

SAS Output

```
              Analysis Of Maximum Likelihood Parameter Estimates

                                   Standard   Wald 95% Confidence          Wald
```

```
Parameter   DF   Estimate      Error          Limits          Chi-Square

Intercept    1     2.0517     0.0383     1.9767     2.1267      2874.02
Scale        0     1.0000     0.0000     1.0000     1.0000

                Analysis Of Maximum Likelihood Parameter Estimates

                          Parameter    Pr > ChiSq

                          Intercept       <.0001
                          Scale

                NOTE: The scale parameter was held fixed.


         Analysis Of Maximum Likelihood Zero Inflation Parameter Estimates

                                   Standard   Wald 95% Confidence
Parameter   DF   Estimate           Error          Limits

Intercept    1    -0.2298          0.1602    -0.5438     0.0843
```

Note that in the SAS output under the title *Analysis Of Maximum Likelihood Zero Inflation Parameter Estimates*, the *Estimate*$=-0.2298$ is the estimate of the inflation parameter p. In the GENMOD code, since we used *zeromodel /link=logit*, the inflation parameter p is estimated in logit scale by reparametrizing $p = (1 - 1/exp(\phi))$, where ϕ is the intercept. Thus the MLE of the inflation parameter p is

$$\begin{aligned}\widehat{p} &= \frac{1}{(1 + exp(-(-0.2298)))} \\ &= 0.4428.\end{aligned}$$

From the confidence interval of $\widehat{\phi}$ above, inverting back the logit scale, we find that the approximate 95% confidence interval of p is $[0.3673, 0.5210]$. Thus, p is significantly different from zero.

Similarly, in the SAS output under the title *Analysis Of Maximum Likelihood Parameter Estimates*, the *Estimate=2.0517* corresponding to *Intercept* is the estimate of the parameter λ. These estimates are computed using *log* scale (link=log is the default link of model statement in GENMOD). Hence, inverting back the log-transformed estimate $\widehat{\lambda}$ to the raw scale, the estimate of $\widehat{\lambda}$ will be $exp(2.0517) = 7.7811$. Thus the MLE of the mean is

$$\begin{aligned}\widehat{\mu} &= \widehat{\lambda}\widehat{\pi} \\ &= \widehat{\lambda}(1 - \widehat{p}) \\ &= 7.7811 \times (1 - 0.4428) \\ &= 4.3356\end{aligned}$$

and the MLE of variance is

$$\begin{aligned}\widehat{V} &= \widehat{\lambda}\widehat{\pi} + \widehat{\lambda}^2\widehat{\pi}(1-\widehat{\pi}) \\ &= 7.7811 \times (1-0.4428) + 7.7811^2 \times (1-0.4428) \times 0.4428 \\ &= 19.274\end{aligned}$$

Hence, a 95% confidence interval of μ can be computed as

$$\widehat{\mu} \pm 1.96 \times \sqrt{\widehat{V}/n} \quad = \quad 4.3356 \pm 1.96 \times \sqrt{19.274/158}, \qquad (7.22)$$

which reduces to $[3.6456, 5.0253]$.

Note that PROC GENMOD does not directly provide the estimate of μ, the estimate of its standard error, and the confidence interval for μ. On the other hand, PROC NLMIXED computes these quantities directly.

SAS code:

```
data Tumor;
input Tumor count @@;
cards;
0 70 1 13 2 15 3 6 4 7 5 5
6 9 7 9 8 1 9 2 10 5 11 1 13 1
14 1 15 1 16 1 20 3 21 1
24 3 26 1 30 1 50 2
;
run;
data one; set tumor;
  do i=1 to count;
    y=tumor;
    output;
  end;
drop i;
run;

proc nlmixed data=one;
        /* define zero inflation model */
        zeromodel = a0;
        infprob  = 1/(1+exp(-zeromodel));

        /* define poisson model */
        lambda   = exp(b0);
        /* define ZIP log likelihood */
        if y=0 then ll = log(infprob + (1-infprob)*exp(-lambda));
        else ll = log((1-infprob)) + y*log(lambda) - lgamma(y+1) - lambda;
        model y ~ general(ll);

        /* request mean rate and CI */
estimate "Mean Rate" (1-infprob)*lambda;
run;
```

Output: SAS NLMIXED

```
                    Parameter Estimates

          Parameter      Upper     Gradient

          a0           0.08670    -5.38E-7
          b0            2.1273    2.994E-6
```

Additional Estimates

Label	Estimate	Standard Error	DF	t Value	Pr > |t|	Alpha	Lower	Upper
Mean Rate	4.3354	0.3493	158	12.41	<.0001	0.05	3.6456	5.0253

Note that the 95% confidence interval of μ is $[3.6456, 5.0253]$, which incidentally came out to be exactly the same as the above. In fact, NLMIXED computes the standard error of $\widehat{\mu}$ using Fisher information matrix, which may not always be equal to the standard error estimate using normality assumption $\sqrt{\widehat{V}/n}$.

Furthermore, the confidence intervals based on the models **NBD** and **GPD** are much wider than the interval based on the model **ZIPD**, while both Clopper–Pearson and Mid-p intervals based on Poisson distribution are shorter. The explanation for this could be, the tumor data show both overdispersion and excess zero count. The **ZIPD** could account only for excess zero count but fails to capture the overdispersion.

7.4.4 Zero-inflated generalized Poisson distribution (ZIGPD)

The model is a mixture of a point mass at 0 with probability p and a GPD(λ, ξ) with probability $(1-p)$. Therefore, the probability distribution $f(y|\lambda, \xi)$ of the count Y is given by

$$Pr(Y = y) = \begin{cases} p + (1-p)exp(-\lambda) & y = 0 \\ (1-p)\frac{\lambda(\lambda+y\xi)^{y-1}e^{-(\lambda+y\xi)}}{y!} & y > 0. \end{cases} \quad (7.23)$$

for $\lambda > 0$, and $-1 < \xi < 1$ such that $f(y|\lambda, \xi) = 0$ for $y \geq m$ if $\lambda + m\xi \leq 0$. For the above model, the mean (μ) of Y is

$$\mu_{gp}\pi = (1-p)\mu_{gp}$$

where $\mu_{gp} = \frac{\lambda}{1-\xi}$ is the mean of **GPD**; and the variance (V) is

$$V_{gp}\pi + \mu_{gp}^2\pi(1-\pi) = (1-p)\mu_{gp}\left[p\mu_{gp} + \frac{1}{(1-\xi)^2}\right]$$

where $V_{gp} = \frac{\lambda}{(1-\xi)^3}$ is the variance of **GPD**. If $p = 0$, then the model reduces to GPD.

Illustration: Confidence intervals for zero-inflated generalized Poisson distribution (ZIGPD)

In the following, the computations of the confidence interval of the mean and variance of **ZIGPD** are illustrated using SAS's PROC NLMIXED procedure. We used tumor data introduced earlier containing excess zero count of 44%.

SAS Code

```
data Tumor;
input Tumor count @@;
cards;
0 70 1 13 2 15 3 6 4 7 5 5
6 9 7 9 8 1 9 2 10 5 11 1 13 1
14 1 15 1 16 1 20 3 21 1
24 3 26 1 30 1 50 2
;
run;
data one; set tumor;
  do i=1 to count;
    y=tumor;
    output;
  end;
drop i;
run;

proc nlmixed data=one;
parms _mu_=5 _xi_=0.1 ;
        /* define zero inflation model */
        zeromodel = _phi_;
        infprob  = 1/(1+exp(-zeromodel));
        /*-----xi is a logit transformation --------------*/
xi= 1/(1+exp(-_xi_));
           if xi>=1 then xi=0.99;
        /* Since _mu_=lambda/(1-xi) then lambda   = _mu_*(1-xi)*/
        lambda   = _mu_*(1-xi);
        mustar = _mu_ - xi*(_mu_ - y);
        /* define ZIGP log likelihood */
        if y=0 then ll = log(infprob + (1-infprob)*exp(-lambda));
        else ll = log((1-infprob)) + log(_mu_*(1-xi)) + (y-1)*log(mustar)-mustar
        -lgamma(y+1);
        model y ~ general(ll);

        /*---------estimate mean rate, corresponding CI and variance */
        estimate "Mean Rate" (1-infprob)*_mu_;
        estimate "Variance" (1-infprob)*_mu_*(infprob*_mu_+1/(1-xi)**2);
run;
```

Output

Fit Statistics

-2 Log Likelihood	736.9
AIC (smaller is better)	742.9
AICC (smaller is better)	743.1
BIC (smaller is better)	752.1

Parameter Estimates

Parameter	Estimate	Standard Error	DF	t Value	Pr > |t|	95% Confidence Limits	
_mu_	6.6479	0.9872	158	6.73	<.0001	4.6982	8.5977
_xi_	0.8979	0.2057	158	4.37	<.0001	0.4917	1.3042

```
_phi_       -0.6285    0.2572   158     -2.44     0.0156   -1.1365    -0.1205

                             Additional Estimates

                          Standard
Label         Estimate       Error     DF  t Value  Pr > |t|   Alpha      Lower          Uppe

Mean Rate       4.3354      0.6252    158     6.93    <.0001    0.05     3.1006        5.5703
Variance       61.7636     22.0506    158     2.80    0.0057    0.05    18.2117        105.32
```

In the PROC NLMIXED code, the inflation parameter p is reparametrized in logit scale to avoid computational error, and named as _phi_ (ϕ). From the above SAS output, the MLE estimate $\widehat{\phi}$ is shown as -0.6285. Hence the MLE of the inflation parameter estimate p can be computed as

$$\begin{aligned}\widehat{p} &= \frac{1}{(1+exp(-(-0.6285)))} \\ &= 0.3478.\end{aligned}$$

Similarly, from the estimate of the confidence interval of ϕ, by inverting the logit scale, one can find the approximate 95% confidence interval of p as $[0.2430, 0.4700]$. Since the interval does not include 0, one can conclude that p is significantly different from 0.

Also, in the PROC NLMIXED code, the following statement computes the estimates and the corresponding confidence intervals of the mean and the variance directly.

```
estimate "Mean Rate" (1-infprob)*_mu_;
estimate "Variance" (1-infprob)*_mu_*(infprob*_mu_+1/(1-xi)**2);
```

As shown in the SAS output under *Additional Estimates*, the estimates of the mean and the corresponding 95% confidence interval are reported as 4.3354 and $[3.1, 5.57]$, and the variance estimate and the corresponding 95% confidence interval are computed as 61.7636 and $[18.21, 105.32]$, respectively. Clearly, there is excess variation as the variance is significantly larger than the mean. Also, note that the confidence interval of the mean based on **ZIGPD** is shorter than the intervals based on **NBD** and **GPD**, but wider than the interval based on **ZIP**. The explanation could be that **ZIGPD** takes a part of excess variation into account through the point mass at zero and the remaining through the dispersion parameter.

7.4.5 Zero-inflated negative binomial (ZINB)

The model is a mixture of a point mass at 0 with probability p and a negative binomial distribution with probability $1-p$. The distribution of count Y is given by

$$Pr(Y=y) = \begin{cases} p+(1-p)(\frac{1}{1+d\mu})^{d^{-1}} & y=0 \\ (1-p)\left(\frac{\Gamma(y+d^{-1})}{y!\Gamma(d^{-1})}\right)\left(\frac{d\mu}{1+d\mu}\right)^{y}\left(\frac{1}{1+d\mu}\right)^{d^{-1}} & y>0 \end{cases} \tag{7.24}$$

The mean and variance of ZINB are $\mu_{nb}\pi$ and $V_{nb}\pi + \mu_{nb}^2\pi(1-\pi)$, respectively, where $\mu_{nb} = \mu$ and $V_{nb} = \mu(1 + d\mu)$ are the mean and variance of **NBD**, and $\pi = 1 - p$. The variance V of the ZINB can be further simplified as $V(Y) = (1-p)\,[\mu(1 + p\mu + d\mu)]$. Note that ZINB converges to ZIP when $d \to \infty$. If both $d \to \infty$ and $p \to 0$, then ZINB reduces to Poisson.

Illustration: Confidence intervals for zero-inflated negative binomial distribution (ZINB)

As before, we illustrate the computation of the confidence interval of the mean utilizing the ZINB model for tumor data.

SAS Code: Using PROC GENMOD

```
data Tumor;
input Tumor count @@;
cards;
0 70 1 13 2 15 3 6 4 7 5 5
6 9 7 9 8 1 9 2 10 5 11 1 13 1
14 1 15 1 16 1 20 3 21 1
24 3 26 1 30 1 50 2
;
run;

proc genmod data=tumor;
   model tumor= /dist=zinb;
   zeromodel /link=logit;
   freq count;
   /*output out=_o pzero=pz;*/
run;
```

Output

```
            Analysis Of Maximum Likelihood Parameter Estimates

                                 Standard   Wald 95% Confidence          Wald
 Parameter    DF    Estimate       Error          Limits             Chi-Square

 Intercept     1      1.7439      0.1968      1.3582       2.1297         78.52
 Dispersion    1      1.8427      0.6804      0.8936       3.7997

                    Analysis Of Maximum Likelihood Parameter Estimates

                          Parameter     Pr > ChiSq

                          Intercept        <.0001
                          Dispersion

NOTE: The negative binomial dispersion parameter was estimated by maximum
      likelihood.

      Analysis Of Maximum Likelihood Zero Inflation Parameter Estimates

                                 Standard   Wald 95% Confidence
  Parameter    DF    Estimate      Error          Limits

  Intercept     1     -1.1416     0.6389     -2.3937       0.1105
```

In the above SAS output, under the heading *Analysis of Maximum Likelihood Zero Inflation Parameter Estimates*, the estimate of the inflation parameter p is shown in logit scale. This is because in the SAS code, we used *zeromodel /link=logit*. The MLE of the parameter p in the logit scale is shown as -1.1416 (the estimate for the parameter intercept). Hence, the MLE of the inflation parameter p is

$$\begin{aligned}\widehat{p} &= \frac{1}{(1+exp(-(-1.1416)))} \\ &= 0.2403.\end{aligned}$$

Similarly, from the confidence interval of ϕ, by inverting the logit scale, one can find the approximate 95% confidence interval of p as $[0.0836, 0.5276]$.

The MLE of μ is $exp(1.7439) = 5.7196$. Hence, the estimated mean from the model is

$$\begin{aligned}&\widehat{\mu\pi} \\ &= \widehat{\mu}(1-\widehat{p}) \\ &= 5.7196 \times (1 - 0.242) \\ &= 4.3356\end{aligned}$$

and the estimated variance from the model is

$$\begin{aligned}\widehat{V} &= \widehat{\mu}(1+\widehat{d\mu})\widehat{\pi} + \widehat{\mu}^2\widehat{\pi}(1-\widehat{\pi}) \\ &= \widehat{\pi}\left[\widehat{\mu}\left(1+\widehat{d\mu}+\widehat{\mu}(1-\widehat{\pi})\right)\right] \\ &= (1-\widehat{p})\left[\widehat{\mu}(1+\widehat{p\mu}+\hat{d\mu})\right] \\ &\approx 56.0327\end{aligned}$$

The above estimate of the mean and variance can be verified directly using the following SAS code using *PROC FMM*.

```
proc fmm data=tumor noprint;
  model tumor= /dist=nb;
  model + /dist=constant;
  freq count;
output out=out1 pred=mean variance=Var;
run;
proc print data=out1 noobs; run;
```

SAS shows the following output (first four observations are printed below):

```
Tumor    count      mean        Var

  0        70     4.33544    56.0325
  1        13     4.33544    56.0325
  2        15     4.33544    56.0325
  3         6     4.33544    56.0325
  4         7     4.33544    56.0325
 ...       ..                 ....
```

Using normality assumption, the MLE based 95% confidence interval of μ can be computed as

$$\widehat{\mu\pi} \pm 1.96 \times \sqrt{\widehat{V}/n} \quad = \quad 4.3356 \pm 1.96 \times \sqrt{56.0325/158} \qquad (7.25)$$

which reduces to $(3.1684, 5.5028)$.

As stated above, even though PROC GENMOD can compute the mean, but cannot directly compute the confidence interval of the mean. Similar to section 7.4.3.1, PROC NLMIXED can be used to estimate the mean and the corresponding confidence interval. The following SAS code can be used:

```
data Tumor;
input Tumor count @@;
cards;
0 70 1 13 2 15 3 6 4 7 5 5
6 9 7 9 8 1 9 2 10 5 11 1 13 1
14 1 15 1 16 1 20 3 21 1
24 3 26 1 30 1 50 2
;
run;

data one; set tumor;
  do i=1 to count;
    y=tumor;
    output;
  end;
drop i;
run;

proc nlmixed data=one;
        /* define zero inflation model */
        zeromodel = _phi_;
        infprob  = 1/(1+exp(-zeromodel));

        /* define poisson model */
        mu    = exp(intercept);

        /* define ZINB log likelihood */

if y=0 then ll = log( infprob + (1-infprob)/(1+d*mu)**(1/d) );
        else ll = log((1-infprob)) + y*log(d*mu) - (y+(1/d))*log(1+d*mu) +
        lgamma(y+1/d) - lgamma(1/d) - lgamma(y+1);

        model y ~ general(ll);

        /* request mean/variance and corresponding CI */
estimate "Mean Rate" (1-infprob)*mu;
estimate "Variance" (1-infprob)*mu*(1+infprob*mu+mu*d);
run;
```

SAS shows the following output.

```
                              Parameter Estimates

                          Standard                                    95% Confidence
Parameter   Estimate        Error     DF    t Value   Pr > |t|          Limits

_phi_        -1.1416       0.6388    158      -1.79     0.0759    -2.4033     0.1202
```

```
intercept     1.7439    0.1968   158      8.86    <.0001    1.3552    2.1326
d             1.8427    0.6804   158      2.71    0.0075    0.4989    3.1865

                          Additional Estimates

                      Standard
 Label       Estimate    Error   DF t Value Pr > |t| Alpha     Lower     Upper

 Mean Rate    4.3355    0.5955  158     7.28   <.0001  0.05    3.1593    5.5117
 Variance    56.0327   16.4530  158     3.41   0.0008  0.05   23.5364   88.5290
```

In the SAS output, under title *Additional Estimates*, the MLE of mean and its confidence interval are shown as 4.3355 and $(3.1593, 5.5117)$, respectively. Note that the confidence interval is slightly different from the interval estimated based on normality assumption using the estimates of PROC GENMOD. The difference arises because, in the PROC NLMIXED interval, the estimate of the standard error of $\widehat{\pi}\widehat{\mu}$ is computed from the Fisher information matrix, while in GENMOD, it is approximated by $\widehat{V}/n$.

7.5 Bayesian credible intervals for rate parameter

This section briefly describes the method to obtain credible intervals for Poisson mean and includes the SAS code to carry out the computation. Using Bayes theorem, given the likelihood $P(\boldsymbol{y}|\boldsymbol{\theta})$ and the prior $P(\boldsymbol{\theta})$, the posterior distribution of $\boldsymbol{\theta}$ is:

$$P(\boldsymbol{\theta}|\boldsymbol{y}) \propto P(\boldsymbol{y}|\boldsymbol{\theta})P(\boldsymbol{\theta}) \tag{7.26}$$

In the following, for illustrative purposes, we consider only flat prior. For more sophisticated Bayesian methods, we recommend [85] and [41].

Poisson distribution Let $y_1, y_2, ...$ represent a random sample from the Poisson distribution with mean μ. Then the likelihood for the mean parameter μ is

$$f(y_i|\mu)) \propto \frac{e^{-\mu}(\mu)^{y_i}}{y_i!} \tag{7.27}$$

where $f(.|.)$ is the conditional density. By equation 7.26, the posterior corresponding to Poisson distribution is

$$P(\mu|\boldsymbol{y}) \propto P(\boldsymbol{y}|\mu)P(\mu)$$

where $P(\mu)$ is the prior distribution of μ. In the following examples, the prior distribution is assumed to be a non-informative or a flat prior.

Illustration: Bayesian credible intervals for Poisson distribution
We illustrate the computation of Bayesian credible intervals of μ for the tumor data using PROC MCMC.

```
data Tumor;
input Tumor count @@;
cards;
0 70 1 13 2 15 3 6 4 7 5 5
6 9 7 9 8 1 9 2 10 5 11 1 13 1
14 1 15 1 16 1 20 3 21 1
24 3 26 1 30 1 50 2
;
run;

data one; set tumor;
  do i=1 to count;
    y=tumor;
    output;
  end;
drop i;
run;

ods graphics on;
/* propcov=quanew initialize chain at the posterior mode and uses the
 estimated inverse Hessian matrix as the initial proposal covariance matrix  */
   proc mcmc data=one seed=12345 nmc=100000 thin=10
      propcov=quanew monitor =(_parms_ Pearson mean);
      parms logmu 0; /*logarithm of mu, computations will be in log scale */
      prior logmu ~ normal(0,var=1000);
      mu = exp(logmu);
      model y ~ poisson(mu);
beginnodata;
        mean=mu;
      endnodata;
   run;
   ods graphics off;
```

The above SAS code is very intuitive. The SAS option *ods graphics on* enables SAS graphics. Like any other SAS procedures, the statement *model y poisson(mu)* is for modeling of the outcome y, where y is assumed to be the Poisson distribution with parameter *mu* and *mu = exp(logmu)*. The statement *prior logmu* $\sim$ *normal(0,var=1000)* specifies a flat prior *normal(0,var=1000)* for the parameter *logmu*. A part of the SAS output is given below, and some of the diagnostic plots are shown in figure 7.2.

Posterior Summaries and Intervals

Parameter	N	Mean	Standard Deviation	95% HPD Interval	
logmu	10000	1.4657	0.0378	1.3933	1.5413
mean	10000	4.3337	0.1636	4.0154	4.6566

Note that the posterior mean and the corresponding credible interval (95% HPD Interval) are reported as 4.3337 and $(4.0154, 4.6566)$. These numbers are slightly different than the earlier results from frequentist methods.

Zero-inflated Poisson distribution As stated in section 7.4.3.1, ZIPD is a mixture of a point mass at 0 with probability p and a Poisson(μ) distribution

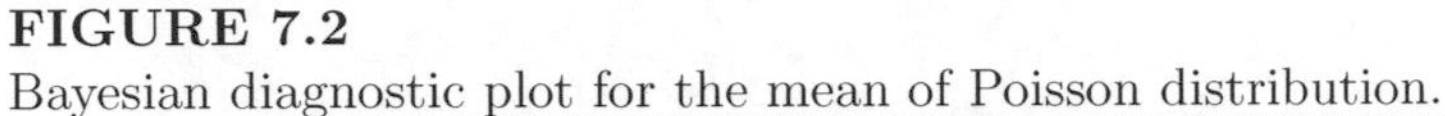

FIGURE 7.2
Bayesian diagnostic plot for the mean of Poisson distribution.

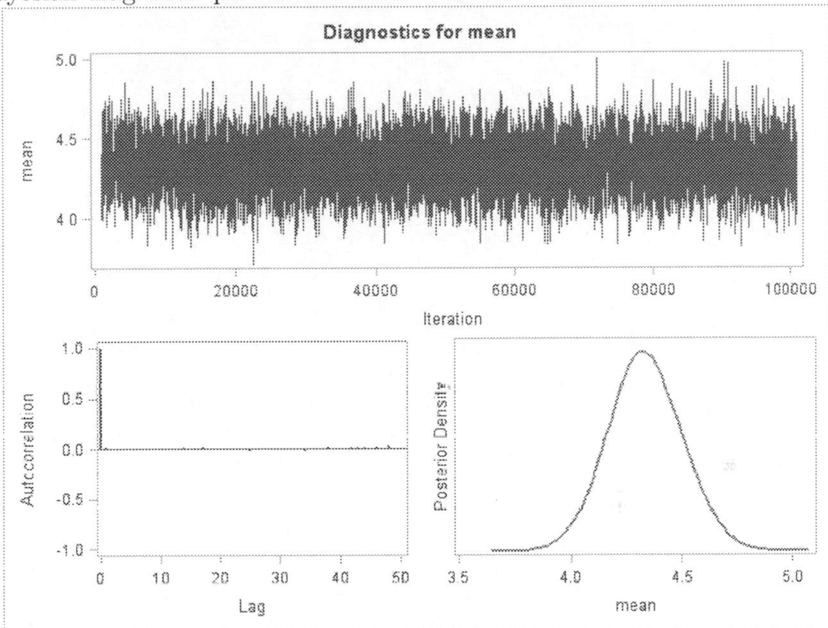

with probability $(1-p)$. Hence, the mean and variance of Y are $E(Y) = \mu\pi$ and $V = V(Y) = \mu\pi + \mu^2\pi(1-\pi)$, respectively, where $\pi = 1-p$. Here $\boldsymbol{\theta} = (\mu, p)$ is a vector-valued parameter. For the prior distribution of $\boldsymbol{\theta}$, we assume $P(\boldsymbol{\theta}) = P(\mu) \times P(p)$. Thus the posterior distribution of $\boldsymbol{\theta}$ becomes

$$P(\mu, p|\boldsymbol{y}) \propto P(\boldsymbol{y}|\mu, p) \times P(\mu) \times P(p),$$

where $P(\boldsymbol{y}|\mu, p)$ is the likelihood function based on **ZIPD**.

Illustration: HPD intervals for zero-inflated Poisson distribution

The MCMC diagnostic plots are shown in Figure 7.2 show the convergence of the chain to a stationary distribution. The posterior distribution shown here helps in computing the Bayesian credible interval with tumor data.

```
data Tumor;
input Tumor count @@;
cards;
0 70 1 13 2 15 3 6 4 7 5 5
6 9 7 9 8 1 9 2 10 5 11 1 13 1
14 1 15 1 16 1 20 3 21 1
24 3 26 1 30 1 50 2
;
run;
```

```
data one; set tumor;
  do i=1 to count;
    y=tumor;
    output;
  end;
drop i;
run;

    ods graphics on;
   proc mcmc data=one seed =12345 nmc=100000 thin=10
      propcov=quanew monitor =(_parms_ Pearson mean sigma2);
/*      ods select Parameters PostSummaries PostIntervals tadpanel;*/
      parms logmu 0 _p_ .3;
      prior logmu ~ normal(0,var=1000);
      prior _p_ ~ uniform(0,1);
      mu=exp(logmu);
      llike=log(_p_*(y eq 0) + (1-_p_)*pdf("poisson",y,mu));
      model general(llike);
        beginnodata;
           mean = (1 - _p_)*mu;
           sigma2 = (1 - _p_)*mu*(1 + _p_*mu);
        endnodata;
   run;
   ods graphics off;
```

In the above SAS code, *_p_* is for the zero-inflation parameter p, and *logmu* is coded for the log transformed value of the mean of the Poisson distribution μ. The priors specified for *_p_* and *logmu* are uniform$(0,1)$ and normal$(0,1000)$, respectively. The statement "*llike=log(_p_*(y eq 0) + (1–_p_)*pdf("poisson",y,mu))*" is used for the construction of the log-likelihood. Finally, the statement "*model general(llike)*" specifies a general log-likelihood function that one constructs using SAS programming statements. By specifying expressions *mean* and *sigma2* inside the block *beginnodata* and *endnodata*, one can compute the posterior mean and the corresponding HPD intervals of the *mean*, and the estimate of the variance *sigma2* and the corresponding HPD confidence intervals of the variance. The following is a part of the SAS output:

Posterior Summaries and Intervals

Parameter	N	Mean	Standard Deviation	95% HPD Interval	
Logmu	10000	2.0507	0.0382	1.9776	2.1266
_p_	10000	0.4439	0.0390	0.3695	0.5218
mean	10000	4.3261	0.3452	3.6693	5.0102
sigma2	10000	19.1947	1.3076	16.7586	21.8781

Note that the posterior mean of p (in the output *_p_*)and its 95% HPD interval are given by 0.4439 and $[0.3695, 0.5218]$, which clearly indicates significant excess occurrence of zero count. The posterior mean of the expression $(1-p)\lambda$ and its 95% HPD interval are given by 4.3261 and $[3.6693, 5.0102]$. As expected, the interval is wider because the distribution captures the extra variability. The posterior mean of the variance parameter is estimated as (19.1947) compared to 4.3337 for Poisson. SAS output also displays diagnostic plots similar to Poisson distribution example discussed earlier, but the figures are not included here.

The posterior mean and the HPD interval for a parameter of a distribution such as **NBD**, **GPD**, **ZINBD** or **ZIGPD** can be computed similarly. In the PROC MCMC code, one needs to declare the parameters of the model, corresponding prior, and the likelihood of the distribution. The SAS codes discussed above provide an outline of PROC MCMC, but we refer the interested readers to PROC MCMC documentation of SAS for further details. In this connection, note that in SAS, there are a number of SAS procedures such as GENMOD, GLIMMIX that can also provide posterior mean and the HPD interval. Also, OpenBUGS and R provide powerful tools for Bayesian analysis, the use of which is not illustrated here.

7.6 Discussion and recommendation

In this chapter, the focus of our discussion is confidence intervals for the mean of count data distributions. For Poisson distribution, we discuss exact and asymptotic intervals. Exact intervals are usually conservative, and the attained coverage probabilities could be substantially higher than the nominal coverage probability, especially for smaller values of the mean λ. On the other hand, the attained coverage could be significantly lower for asymptotic intervals than the stipulated coverage, particularly for smaller values of λ. Consequently, the average lengths of the exact intervals are more significant than the asymptotic intervals. Which interval to choose in a given context is a matter of personal preference. Among the exact intervals, Blaker's and Casella–Robert–Kabaila–Byrne's are preferred to Clopper–Pearson as the average lengths of the former are shorter. Among the asymptotic intervals, the likelihood ratio interval is the most preferred, followed by the score intervals. Jeffreys' interval is also considered a good choice by Brown et al. ([16]).

We have noted that **NBD** is a model for overdispersed data, while **GPD** could be used to model both over and under-dispersed data. As stated earlier in subsection 7.4.2, for overdispersed data, **GPD** has heavier tail than the **NBD**. In our experience, in most cases, both approaches give similar results. However, it may be prudent to find the confidence interval of the mean using both the models and compare. If there is a significant difference between the two intervals, one could check the goodness of fit of the models and make a decision accordingly.

For **NBD**, Shilane et al. [83] note that the standard intervals (Wald and bootstrap) for mean are typically too narrow and lead to significant under-coverage in the presence of high overdispersion, and they propose some alternative confidence intervals for overdispersion. For **GPD**, we did not find a comparable study. It may be interesting to investigate the performances of

Wald and bootstrap intervals of the mean based on **GPD**, and possibly to develop better intervals.

For data showing only excess zero count, we recommend the **ZIPD** method. On the other hand, for data showing both excess zero count and over/under dispersion, we prefer **ZIGPD** and **ZINB** models. We have discussed both likelihood-based and moment-based Wald type intervals. We expect likelihood-based intervals to perform well if the model is true. However, there is no clear evidence for preferring one to the other. The fact of the matter is that there is little research to guide us on choosing a confidence interval for such models.

Bibliography

[1] A. Agresti and B. Caffo. On small-sample confidence intervals for parameters in discrete distributions. *The American Statistician*, 54:280–288, 2000.

[2] A. Agresti, and Y. Min. On small-sample confidence intervals for parameters in discrete distributions. *Biometrics*, 57:963–971, 2001.

[3] A. Agresti, and B. A. Coull. Approximate is better than “exact” for interval estimation of binomial proportions. *The American Statistician*, 52(2):119–126, 1998.

[4] A. Agresti, and A. Gottard. Nonconservative exact small-sample inference for discrete data. *Computational Statistics & Data analysis*, 51(12):6447–6458, 2007.

[5] A. Agresti, and Y. Min. Simple improved confidence intervals for comparing matched proportions. *Statistics in Medicine*, 24(5):729–740, 2005.

[6] Y. Benjamini. Selective Inference: The Silent Killer of Replicability. Harvard Data Science Review, 2(4). https://doi.org/10.1162/99608f92.fc62b261, 2020.

[7] D.A. Berry. P-values are not what they're cracked up to be. online discussion: ASA statement on statistical significance and p-values. *The American Statistician*, 70(2):1–2, 2016.

[8] D. A. Berry. P-values are not what they're cracked up to be. *Online commentary to [6].(doi: 10.1080/00031305.2016. 1154108)*, 2016.

[9] P. J. Bickel, K. Doksum, and J. L. Hodges. *A Festschrift for Erich L. Lehmann.* CRC Press, 1982.

[10] A. Birnbaum. Confidence curves: An omnibus technique for estimation and testing statistical hypotheses. *Journal of the American Statistical Association*, 56(294):246–249, 1961.

[11] H. Blaker Confidence curves and improved exact confidence intervals for discrete distributions. *Canadian Journal of Statistics*, 28(4):783–798, 2000.

[12] H. Blaker, and E. Spjotvoll. Paradoxes and improvements in interval estimation. *The American Statistician*, 54(4):242–247, 2000.

[13] C. Blyth and H. Still. Binomial Confidence Intervals. *The American Statistician*, 78:108–116, 1983.

[14] L. D. Brown, T. T. Cai, and A. DasGupta. Interval estimation for a binomial proportion. *Statistical Science*, pages 101–117, 2001.

[15] L. D. Brown, T. T. Cai, and A. DasGupta. Confidence intervals for a binomial proportion and asymptotic expansion. *Annals of Statistics*, 30:160–201, 2002.

[16] L. D. Brown, T. T. Cai, and A. DasGupta. Interval estimation in exponential families. *Statistica Sinica*, 13:19–49, 2003.

[17] T. Cai. One-sided confidence intervals in discrete distributions. *Journal of Statistical Planning and Inference*, 131:63–88, 2005.

[18] A. C. Cameron and P. K. Trivedi. Econometric models based on count data: Comparisons and applications of some estimators and tests. *Journal of Applied Econometrics*, 1:29–54, 1986.

[19] R. J. Carrol, and D. Ruppert Robust estimation in heteroscedastic linear models. *Annals of Statistics*, 10:429–441, 1982.

[20] G. Casella. Refining binomial confidence intervals. *Canadian Journal of Statistics*, 14(2):113–129, 1986.

[21] I.S.F. Chan. Exact tests of equivalence and efficacy with a non-zero lower bound for comparative studies. *Statistics in Medicine*, 17:1403–1413, 1998.

[22] I.S.F. Chan, and Z. Zhang. Test based exact confidence intervals for the difference of two binomial proportions. *Biometrics*, 55:1201–1209, 1999.

[23] M. Chang. *Modern Issues and Methods in Biostatistics*. Springer, New York, 2011.

[24] S. C Choi, and D. M. Stablein. Practical tests for comparing two proportions with incomplete data. *Journal of the Royal Statistical Society: Series C (Applied Statistics)*, 31(3):256–262, 1982.

[25] S.-C. Chow, J. Shao, and H. Wang. *Sample Size Calculations in Clinical Research.* 2nd ed. Boca Raton, FL: Chapman & Hall/CRC, 2008.

[26] C. J. Clopper, and E. S. Pearson. The use of confidence or fiducial limits illustrated in the case of the binomial. *Biometrika*, 26:404–413, 1934.

[27] P. R. Coe, and A. C. Tamhane. Small sample confidence intervals for the difference, ratio and odds ratio of two success probabilities. *Communications in Statistics*, 22:925–938, 1993.

[28] Open Science Collaboration et al. Estimating the reproducibility of psychological science. *Science*, 349(6251), 2015.

[29] D. Collett. *Modelling Binary Data.* Chapman & Hall, London, UK, 380, 1991.

[30] P. C. Consul, and G. C. Jain. A generalization of the Poisson distributions. *Technometrics*, 15:791–799, 1973.

[31] R. S. Dann, and G. G. Koch. Review and evaluation of methods for computing confidence intervals for the ratio of two proportions and considerations for noninferiority clinical trials. *Journal of Biopharmaceutical Statistics*, 15:1:85–107, 2004.

[32] A. Donner, and G. Y. Zou. Closed-form confidence intervals for functions of the normal mean and standard deviation. *Statistical Methods in Medical Research*, 21:347–359, 2012.

[33] M. W. Fagerland, S. Lydersen, and P. Laake. Recommended tests and confidence intervals for paired binomial proportions. *Statistics in Medicine*, 33(16):2850–2875, 2014.

[34] M.W. Fagerland, S. Lydersen, and P. Laake. Recommended confidence intervals for two independent binomial proportions. *Statistical Methods in Medical Research*, 24(2):224–254, 2015.

[35] C. P. Farrington, and G. Manning. Test statistics and sample size formulae for comparative binomial trials with null hypothesis of non-zero risk difference or non-unity relative risk. *Statistics in Medicine*, 9:1447–1454, 1990.

[36] B. Fleiss, J. L. Levin, and M. C. Paik. *Statistical Methods for Rates and Proportions.* John Wiley & Sons, Inc., New York, 2003.

[37] CHMP Points to Consider on the Choice of Non-inferiority Margin. CPMP/EWP/2158/99. Published Jul 2005. Effective Jan 2006. Access at http://www.emea.europa.eu/htms/human/humanguidelines/ efficacy.htm

[38] L. F. Fries, S. B. Dillon, J. E. Hildreth, R. A. Karron, A. W. Funkhouser, C. J. Friedman, C. S. Jones, V. G. Culleton, M. L. Clements. Safety and immunogenicity of a recombinant protein inßuenza A vaccine in adult human volunteers and protective efficacy against wild-type H1N1 virus challenge. *Journal of Infectious Diseases*, 167:593–601, 1993.

[39] J. J. Gart, and J. Nam. Approximate interval estimation of the ratio of binomial parameters: A review and correction or skewness. *Biometrics*, 44:323–338, 1988.

[40] F. Garwood. Fiducial limits for the Poisson distribution. *Biometrika*, 28:437–442, 1936.

[41] A. Gelman, J. B. Carlin, H. S. Stern, and D. B. Rubin. *Bayesian Data Analysis*. Chapman and Hall/CRC, 2003.

[42] J. K. Ghosh. On the relation among shortest confidence intervals of different types. *Calcutta Statistical Association Bulletin*, 10(4):147–152, 1961.

[43] S. N. Goodman. Toward evidence-based medical statistics. 1: The p value fallacy. *Annals of Internal Medicine*, 130(12):995–1004, 1999.

[44] S. N. Goodman. The next questions: who, what, when, where, and why. *The American Statistician*, 73(S1), 2016.

[45] M. Greenwood, and G. Yule. An inquiry into the nature of frequency distributions representative of multiple happenings with particular reference to the occurrence of multiple attacks of disease or of repeated accidents. *Journal of the Royal Statistical Society A*, 93:255–279, 1920.

[46] A. Hald. *Statistical Theory with Engineering Applications*. John Wiley and Sons, New York, 1952.

[47] N. A. J. Hastings, and J. B Peacock. *Statistical Distributions*. Wiley, New York, 1975.

[48] H. Joe and R. Zhu. Generalized Poisson distribution: The property of mixture of Poisson and comparison with negative binomial distribution. *Communications in Statistics*, 47(2):219 –229, 2005.

[49] D. Katz, J. Baptista, S. P. Azen, and M. C. Pike. Obtaining confidence intervals for the risk ratio in cohort studies. *Biometrics*, 34:469—474, 1978.

[50] P. A. R. Koopman. Confidence intervals for the ratio of two binomial proportions. *Biometrics*, 40:513–517, 1984.

[51] R. C. Kulkarni, P. M. Tripathi, and J. E. Michalek. Maximum (max) and mid-p confidence intervals and p values for the standardized mortality and incidence ratios. *The American Journal of Epidemiology*, 147:83–86, 1998.

[52] J. F. Lawless. Negative binomial and mixed Poisson regression. *The Canadian Journal of Statistics*, 15(3):209–225, 1987.

[53] J. J. Lee and Z. N. Tu. A versatile one-dimensional distribution plot: the BLiP plot. *The American Statistician*, 51:353–358, 1997.

[54] L. M. Leemis, and K. S. Trivedi. A comparison of approximate interval estimators for the Bernoulli parameter. *The American Statistician*, 50(1):63–68, 1996.

[55] F. D. Liddell. Simple exact analysis of the standardised mortality ratio. *Journal of Epidemiology and Community Health*, 38(1):85–88, 1984.

[56] A. Martin Andres and M. Alvarez Hernandez. Two-tailed approximate confidence intervals for the ratio of proportions. *Statistics and Computing*, DOI 10.1007/s11222–012–9353–5, 2012.

[57] O. S. Miettinen and M. Nurminen. Comparative analysis of two rates. *Statistics in Medicine*, 4:213–226, 1985.

[58] A. M. Millar. Asa statement on p-values: Some implications for education. *The American Statistician*, 70:1–2, 2016.

[59] J. G. Morel and N. K. Neerchal. *Extra Variation Models*. Encyclopedia of Biopharmaceutical Statistics, Third Ed., Editor: Shein-Chung Chow. London: Informa Healthcare, 2010.

[60] R. G. Newcombe. Interval estimation for the difference between independent proportions: Comparison of eleven methods. *Statistics in Medicine*, 17:873–890, 1998.

[61] R. G. Newcombe and M. M. Nurminen. In defence of score intervals for proportions and their differences. *Commun Stat Theory Meth*, 40:1271–1282, 2011.

[62] R. G. Newcombe. Improved confidence intervals for the difference between binomial proportions based on paired data. *Statistics in Medicine*, 17(22):2635–2650, 1998.

[63] R. G. Newcombe. Two-sided confidence intervals for the single proportion: Comparison of seven methods. *Statistics in Medicine*, 17(8):857–872, 1998.

[64] R. G. Newcombe. Measures of location for confidence intervals for proportions. *Communications in Statistics – Theory and Methods*, 40(10):1743–1767, 2011.

[65] R. Nuzzo. Scientific method: Statistical errors. *Nature News*, 506(7487): 150, 2014.

[66] GUIDANCE, DRAFT. "Guidance for industry non-inferiority clinical trials." Center for Biologics Evaluation and Research (CBER) (2010).

[67] M. B. M. Perondi, A. G. Reis, E. F. Paiva, V. M. Nadkarni, and R. Berg. Confidence interval of the difference of two independent binomial proportions using weighted profile likelihood. *New England Journal of Medicine*, 350:1722–1730, 2004.

[68] V. Pradhan and T. Banerjee. Confidence interval of the difference of two independent binomial proportions using weighted profile likelihood. *Communications in Statistics – Simulation and Computation*, 37:645–659, 2008.

[69] V. Pradhan and T. Banerjee. Confidence interval of the difference of two independent binomial proportions using weighted profile likelihood. *Communications in Statistics – Simulation and Computation®*, 37(4):645–659, 2008.

[70] V. Pradhan, J. C. Evans, and T. Banerjee. Binomial confidence intervals for testing non-inferiority or superiority: A practitioner's dilemma. *Statistical Methods in Medical Research*, 25(4):1707–1717, 2016.

[71] V. Pradhan, S. Menon, and U. Das. Corrected profile likelihood confidence interval for binomial paired incomplete data. *Pharmaceutical Statistics*, 12(1):48–58, 2013.

[72] V. Pradhan, K. K. Saha, T. Banerjee, and J. C. Evans. Weighted profile likelihood-based confidence interval for the difference between two proportions with paired binomial data. *Statistics in Medicine*, 33(17):2984–2997, 2014.

[73] J. W. Pratt. Length of confidence intervals. *Journal of the American Statistical Association*, 56(295):549–567, 1961.

[74] R. M. Price and D. G. Bonett. Confidence intervals for a ratio of two independent binomial proportions. *Statistics in Medicine*, 27(26):5497–5508, 2008.

[75] J. Przyborowski and H. Wilenski. Statistical principles of routine work in testing clover seed for dodder. *Biometrika*, 27:273–292, 1935.

[76] G. K. Smyth and M. D. Robinson. Small-sample estimation of negative binomial dispersion, with applications to SAGE data. *Biostatistics*, 9:321–332, 2008.

[77] K. J. Rothman. Disengaging from statistical significance. *European Journal of Epidemiology*, 31(5):443–444, 2016.

[78] K. K. Saha and S. R. Paul. Bias corrected maximum likelihood estimator of the negative binomial dispersion parameter. *Biometrics*, 61:179–185, 2005.

[79] H. Sahai and A. Khurshid. Confidence intervals for the mean of a Poisson distribution: A review. *Biometrical Journal*, 35:857–867, 1993.

[80] T. J. Santner, V. Pradhan, P. Senchaudhuri, C. R. Mehta, and A. Tamhane. Small-sample comparisons of confidence intervals for the difference of two independent binomial proportions. *Computational Statistics and Data Analysis*, 51:5791–5799, 2007.

[81] T. J. Santner and M. K. Snell. Small-sample confidence intervals for $p_1 - p_2$ and p_1/p_2 in 2×2 contingency tables. *Journal of the American Statistical Association*, 75:963–971, 1980.

[82] T. J. Santner and S. Yamagami. Invariant small sample confidence intervals for the difference of two success probabilities. *Communications in Statistics*, 22:33–59, 1993.

[83] D. Shilane, A. E. Hubbard, and S. N. Evans. Confidence Intervals for Negative Binomial Random Variables of High Dispersion. U.C. Berkeley Division of Biostatistics Working Paper Series 242, 2008.

[84] K. Sidik. Exact unconditional tests for testing non-inferiority in matched-pairs design. *Statistics in Medicine*, 22(2):265–278, 2003.

[85] K. Spiegelhalter, D. Abrams and J. Myles. *Bayesian Approaches to Clinical Trials and Health Care Evaluation.* John Wiley and Sons, Chichester, UK, 2004.

[86] P. B. Stark. The value of p-values, `https://s3-eu-west-1.amazonaws.com/pstorage-tf-iopjsd8797887/.../21_Stark.pdf`. 2015.

[87] T. E. Sterne. Some remarks on confidence of fiducial limits. *Biometrika*, 41:275–278, 1954.

[88] G. M. Sullivan and R. Feinn. Using effect size – or why the p value is not enough. *Journal of Graduate Medical Education*, 4(3):279, 2012.

[89] M.-L. Tang, M.-H. Ling, L. Ling, and G. Tian. Confidence intervals for a difference between proportions based on paired data. *Statistics in Medicine*, 29(1):86–96, 2010.

[90] M.-L. Tang, M.-H. Ling, and G.-L. Tian. Exact and approximate unconditional confidence intervals for proportion difference in the presence of incomplete data. *Statistics in Medicine*, 28(4):625–641, 2009.

[91] T. Tango. Equivalence test and confidence interval for the difference in proportions for the paired-sample design. *Statistics in Medicine*, 17(8):891–908, 1998.

[92] M. A. Turco, J. A. Ormiston, J. J. Popma, L. Mandinov, C. D. O′Shaughnessy, T. Mann, T. F. McGarry, C. J. Wu, C. Chan, M. W. I. Webster, J. J. Hall, G. J. Mishkel, L. A. Cannon, D. S. Baim, and J. Koglin. Polymer-based, Paclitaxel-Eluting TAXUS Liberte′ Stent in De Novo Lesions. *Journal of the American College of Cardiology*, 49(16):1676–1683, 2007.

[93] K. Ulm. A simple method to calculate the confidence interval of a standardized mortality ratio. *Journal of Epidemiology and Community Health*, 131(2):373–375, 1984.

[94] D. J. Venzon and S. H. Moolgavkar. A method for computing profile-likelihood based confidence intervals. *Applied Statistics*, 37:87–94, 1998.

[95] S. E. Vollset. Confidence intervals for a binomial proportion. *Statistics in Medicine*, 12(9):809–824, 1993.

[96] S. D. Walter. The distribution of Levin's measure of attributable risk. *Biometrika*, 62(2):371–374, 1975.

[97] R. L. Wasserstein and N. A. Lazar. The ASA statement on *p*-values: context, process, and purpose. *The American Statistician*, 70(2):129–133, 2016.

[98] R. W. M. Wedderburn. Quasi-likelihood function, generalized linear models and the Gauss-Newton method. *Biometrika*, 61:439–447, 1974.

[99] E. B. Wilson. Probable inference, the law of succession, and statistical inference. *Journal of the American Statistical Association*, 22:209–212, 1927.

[100] E. B. Wilson. Probable inference, the law of succession, and statistical inference. *Journal of the American Statistical Association*, 22(158):209–212, 1927.

[101] F. Yates. Contingency tables involving small numbers and the χ^2 test. *Journal of the Royal Statistical Society*, (Suppl. 1):217–235, 1934.

Index

Note: Locators in *italics* represent figures and **bold** indicate tables in the text.

Printed in the United States
by Baker & Taylor Publisher Services